Lipids and Heart Disease

A Practical Approach

MADELEINE BALL

Clinical Lecturer, Nuffield Department of Clinical Biochemistry
John Radcliffe Hospital, Oxford

and

JIM MANN

Professor of Human Nutrition, University of Otago, New Zealand

Oxford New York Tokyo
OXFORD UNIVERSITY PRESS

Oxford University Press, Walton Street, Oxford OX2 6DP
Oxford New York Toronto
Delhi Bombay Calcutta Madras Karachi
Petaling Jaya Singapore Hong Kong Tokyo
Nairobi Dar es Salaam Cape Town
Melbourne Auckland
and associated companies in
Berlin Ibadan

Oxford is a trade mark of Oxford University Press

Published in the United States
by Oxford University Press, New York

First published 1988
Reprinted 1989, 1990, 1991

British Library Cataloguing in Publication Data
Ball, Madeleine
Lipids and heart disease.
1. Man. Heart. Diseases. Role of lipids
I. Title II. Mann, Jim
616.1'2071
ISBN 0-19-261701-X

Library of Congress Cataloging-in-Publication Data
Ball, Madeleine.
Lipids and heart disease.
(Oxford medical publications)
Bibliography: p.
Includes index.
1. Coronary heart disease. 2. Coronary heart disease—Prevention. 3. Hyperlipidemia. I. Mann, Jim. II. Title. III. Series. [DNLM: 1. Coronary Disease—etiology. 2. Coronary Disease—prevention & control. 3. Hyperlipidemia—diagnosis. 4. Hyperlipidemia—therapy. 5. Lipids—adverse effects. WG 300 B187L]
RC685.C6B254 1988 616.1'23071 88-15121
ISBN 0-19-261701-X (pbk.)

Printed in Great Britain by
Dotesios Printers Ltd, Trowbridge, Wiltshire

Foreword

by

Professor Barry Lewis

Professor of Chemical Pathology and Metabolic Disorders, St Thomas' Hospital, London

This short account of hyperlipidaemia and its relation to coronary heart disease is based on the authors' long experience of the management of this group of metabolic disorders. It provides a wealth of information, in an eminently readable form. The assessment of coronary heart disease risk factors is rapidly becoming a part of established medical practice, and a rapidly growing number of patients are requiring proper diagnosis and treatment. Hence this book comes at an opportune time and will do much to facilitate the handling of such patients in general practice. I warmly welcome this short, readable, and informative book. It comes from two of the most distinguished clinical investigators in Britain. I hope it will be widely read; its availability will undoubtedly further the control of the major coronary heart disease risk factor, hyperlipidaemia.

Acknowledgements

We wish to thank the following individuals for their assistance. Dr Peter Pritchard provided the impetus for the book. Three general practitioners, Chris Stretton, M. Graham, and D. Almond read the draft manuscript and provided helpful comments. Dr Iain Robertson wrote the chapter on screening in general practice. Mrs Jane Thompson helped to produce some of the diet tables in chapter 8, and Professor Barry Lewis kindly allowed us to use several of his clinical photographs in chapter 7.

Contents

Glossary

Apolipoproteins (apoproteins)—protein components of lipoproteins that are involved in structure, cell recognition, and the metabolism of the lipoproteins.

Lipoproteins—lipids are insoluble in water, and are therefore transported in plasma in particles containing various proportions of protein, phospholipid, triglyceride, and cholesterol.

VLDL	— very low density lipoprotein	
IDL	— intermediate density lipoprotein	classification based on
LDL	— low density lipoprotein	centrifugation data
HDL	— high density lipoprotein	

Non-esterified fatty acids—single fatty acid molecules. Three of these in combination with glycerol form a triglyceride molecule.

Saturated fatty acids—these are fatty acids with no double bonds in the carbon skeleton. Stearic and palmitic acids are the commonest; high levels of these are found in most hard animal fats.

Monounsaturated fatty acids—fatty acids with a single unsaturated C=C bond e.g. oleic acid. Olive oil is rich in this fatty acid.

Polyunsaturated fatty acids—fatty acids with multiple C=C bonds e.g. linoleic acid. Sunflower seed, soya, and corn oil contain considerable quantities of these fatty acids.

Units—cholesterol 1 mmol/l $\simeq$39 mg/dl; triglyceride 1 mmol/l $\simeq$88 mg/dl.

Introduction

Coronary heart disease (CHD) is a major cause of death in most countries in the Western world. In 1985, over 160 000 people in the U.K. died of CHD. Thirty one per cent of all male deaths are from CHD and each year about 30 000 of the men who die are under the age of 65, and 5000 are under the age of 55. Figure 1 shows the overall causes of death in men and women.

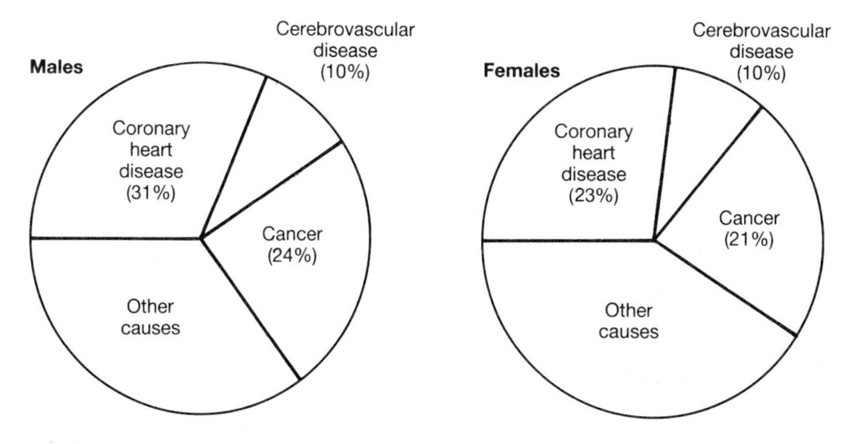

Fig. 1. Overall causes of death in the UK as a percentage of the total (1985).

The British Regional Heart Study showed that by 55–59 nearly one in three men had symptoms or signs of CHD. Although the death rate has been falling in some countries, such as the United States, Finland, New Zealand, and Australia, it remains relatively stable in others, including the UK (Fig. 2), and is rising in some Eastern european countries.

The major risk factors for CHD include smoking, hypertension, and raised blood lipids. Much has been written in the last 10 years about the importance of lipids, particularly cholesterol, in determining CHD risk. The variation in CHD between populations is largely explained by differences in cholesterol levels. The large number of people in the U.K.

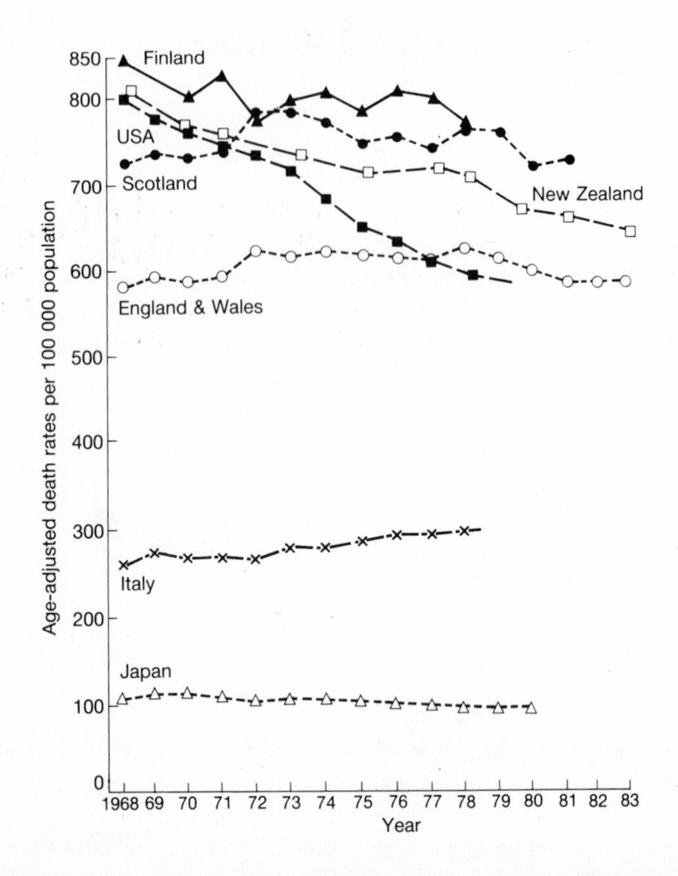

Fig. 2. Trends in CHD mortality. Coronary heart disease mortality 1968–82 in selected countries (men aged 35–74 years). (From Marmot, 1985, *Acta Med. Scand. suppl.*, **701**, 58–65.)

with a plasma cholesterol concentration above 'ideal' can thus account for the high incidence of CHD. In populations with a high CHD risk increased plasma cholesterol contributes to CHD incidence in two ways. First, there are a large number of people (probably half the population) with a slight to moderate increase in plasma lipid levels which result in some increase in risk of CHD. Second, there is a smaller group of people with high levels, often due to an inherited disorder, who are at particularly high risk for premature CHD (Fig. 3).

involved in the retrograde transport of cholesterol from peripheral tissues to the liver.

Hyperlipidaemia

A high concentration of circulating lipoproteins usually results from an increase in their synthesis due to a diet high in saturated fat, and/or a genetically-determined reduction in the removal from the circulation. Depending on the type of particles this causes an increase in the concentration of cholesterol and/or triglyceride in the plasma.

2 Plasma lipids and coronary heart disease

What are normal lipid levels?

Many departments of chemical pathology provide a normal or reference range in brackets after the result of a requested biochemical measurement. We have become accustomed to passing any results which lie within this range as satisfactory. However, it is not always appreciated that the normal range is a statistical concept based upon two standard deviations above and below the mean level of a group of apparently healthy people from the population. For most biochemical measurements the statistical normal range is equivalent to an optimal range, but this does not apply to measurements of blood lipid.

Total plasma cholesterol: relationship to coronary heart disease

No other blood constituent varies so much between different populations as the plasma cholesterol. From New Guinea to East Finland, the mean plasma cholesterol ranges from 2.6 to 7.1 mmol/l (100–275 mg/dl) when estimated by the same method in the same age and sex group.

Relationship between populations

Of the known risk factors for CHD total plasma cholesterol appears to be the most important determinant of the geographical distribution of the disease. This was most clearly demonstrated in the Seven Country Study carried out by Keys and co-workers. These researchers measured actual food consumption and various factors known to be related to CHD (including cigarette smoking, blood-pressure, obesity, and plasma cholesterol) in 16 groups of people living in seven countries and followed each group to determine 10-year incidence rates of CHD. As shown in Figure 2.1, median cholesterol values were highly correlated with CHD death rates ($r = 0.80$) and cholesterol accounted for 64 per cent of the variance in the CHD death rates among the groups. Dietary saturated fat

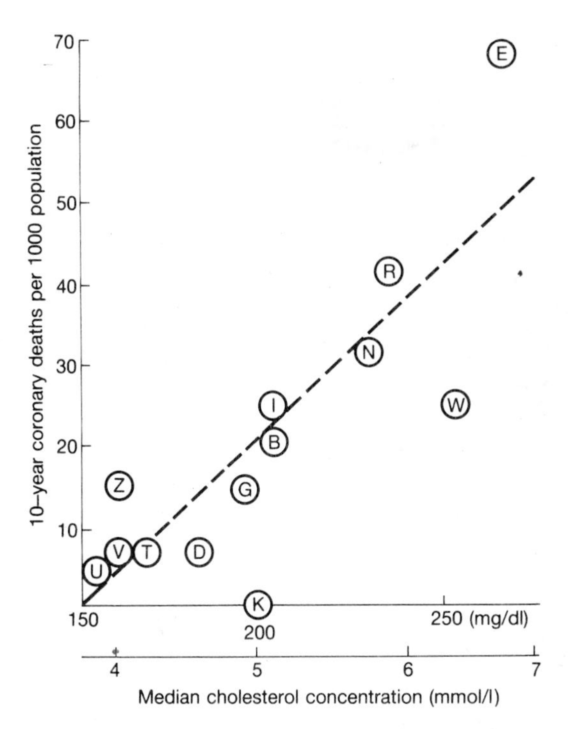

Fig. 2.1. Relationship between the mean cholesterol in different countries and CHD death rates. The relationship of the median plasma cholesterol level to 10-year CHD mortality in 16 male cohorts of the Seven Countries Study. B = Belgrade (Yugoslavia); D = Dalmatia (Yugoslavia); E = East Finland; G = Corfu; I = Italian railroad; K = Crete; N = Zutphen (Holland); T = Tanushimaru (Japan); R = American railroad; U = Ushibuka (Japan); V = Velike Krsna (Yugoslavia); W = West Finland; Z = Zrenjanin (Yugoslavia). (From Keys, A. Seven countries. A multivariate analysis of death and coronary heart disease. Harvard University Press, 1980.)

intake was also strongly correlated with cholesterol and with 10-year CHD mortality. Interestingly, none of the other factors known to be related to CHD (cigarette smoking, blood-pressure, obesity, and physical inactivity) explained the geographical variation of CHD to any appreciable extent. This suggests that the emergence of CHD in a community depends strongly upon the cholesterol level of the community and that the other factors become important only when the population is at risk because of increased cholesterol levels.

Within populations

Among individuals within populations the association between cholesterol and CHD is equally strong: in over 20 prospective studies in different countries total serum cholesterol has been shown to be related to the development of CHD. The association occurs in both sexes and is independent of all other measured risk factors. The relationship in the largest prospective study of over 350 000 people is shown in Figure 2.2. A clear 'dose-related' effect is apparent with a gradient of risk from the lowest to the highest levels. There is no discernible critical value below which there is no risk; the risk tends to increase throughout the range.

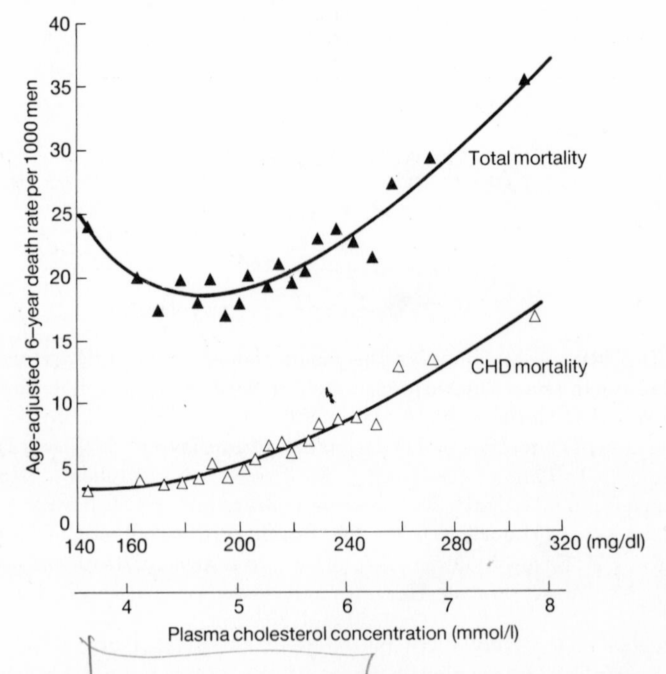

Fig. 2.2. Relationship between plasma cholesterol and CHD—MRFIT Study. (Martin, M. *et al.* (1986). *Lancet*, **2**, 933–6.)

Comparison of groups of people within one population who eat different diets, shows that diet has some effect, although it does not fully explain the lipid levels within populations. Figure 2.3 shows that vegans,

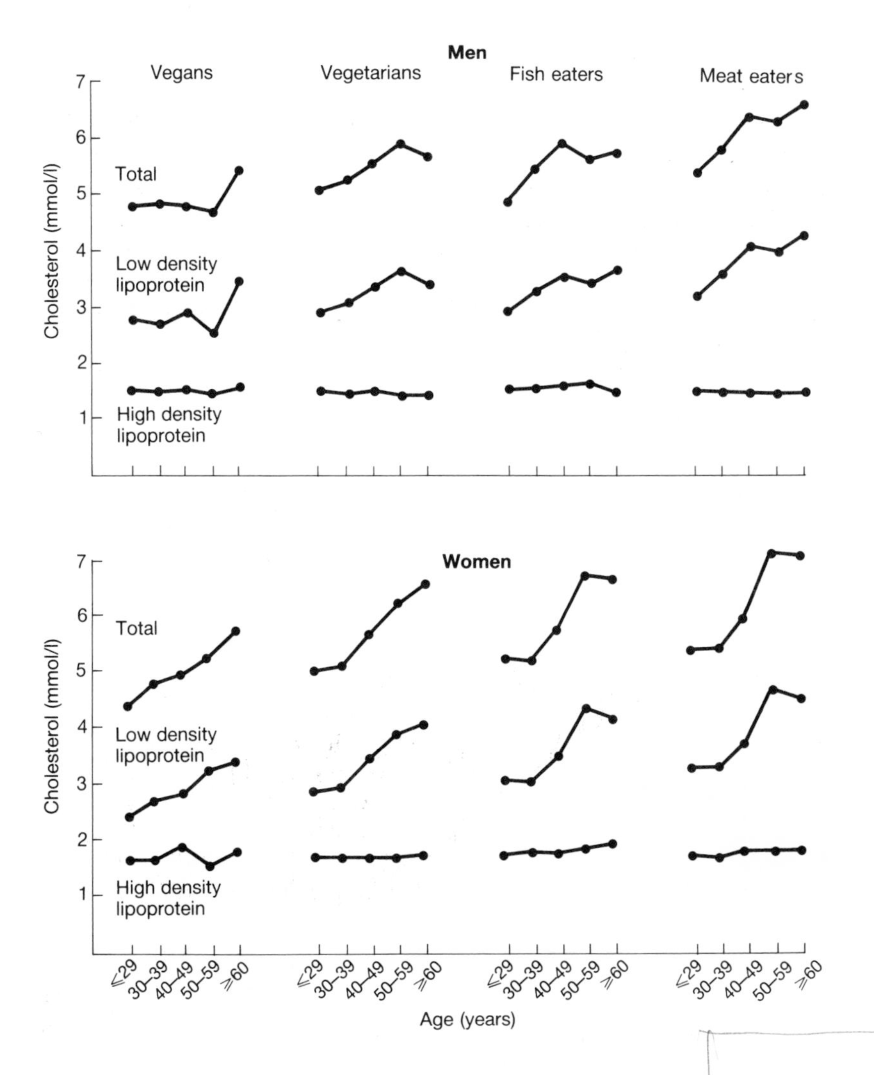

Fig. 2.3. Plasma cholesterol concentration and dietary habits. (Thorogood *et al.* (1987). *British Medical Journal*, **295**, 351–4.)

vegetarians, and those who eat fish, but no meat, have lower levels of cholesterol than meat eaters within the British population. These groups have been found to have lower rates of CHD than meat eaters and this may be explained by the lower cholesterol levels. Diet thus appears to exert a profound effect, explaining both the differences in cholesterol levels and CHD rates among populations and among groups with different dietary practices living within high-risk populations.

Reference range versus optimal level

The distribution of cholesterol levels has been examined in a number of different countries. The 'normal' or reference range (based on two standard deviations above and below the mean), calculated from the cholesterol levels of 25–59-year-olds in the British Lipid Screening Project, is from 3.5 to 8.0 mmol/l. (Fig. 2.4). A similar range is found in countries like New Zealand. It is thus clear that for cholesterol 'normal' certainly does not imply 'optimal'. Those at the upper end of the range have a greater risk of CHD than those at the bottom. Under these circumstances it is clearly desirable to try to define an optimal range. As with other biological variables showing a gradient of risk (such as blood-pressure) this is not an easy task. The European Atherosclerosis Group has suggested that cholesterol levels greater than 5.2 mmol/1 (210 mg/dl) warrant consideration and recommend a graded response to increasing levels. Other official bodies, such as the British Cardiac Society, have recommended that individual advice should be given at levels over 6.5 mmol/1 (255 mg/dl). The proportion of the British population with cholesterol levels greater than these cut-off points is shown in Figure 2.5, and the pros and cons of the use of different cut-off points are discussed in Chapter 11.

Overall cholesterol levels are not appreciably different in men and women, but different age trends are apparent. In the younger age groups, women have lower levels than men. However, there is a steady gradient for both total and LDL cholesterol with increasing age in women, as shown in Figure 2.6, which may partly explain the rapid rise in CHD after the menopause. Levels in men generally increase only until the mid-40s after which they tend to level off. It is this observation that has led some people to suggest that the advisable and 'action limits' for plasma cholesterol should be age and sex adjusted. We do not know if this is sensible because it is not yet clear whether a raised cholesterol confers a different risk at different ages. The increase with age is less marked in populations with lower mean cholesterol levels. HDL levels remain relatively constant with age.

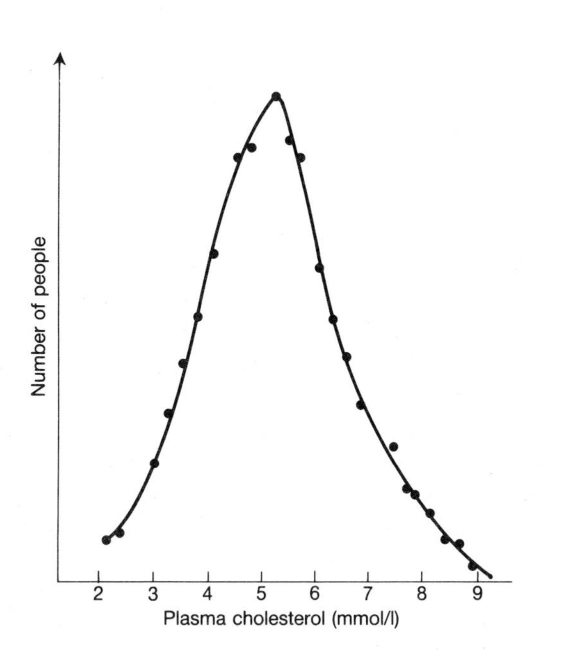

Fig. 2.4. Distribution of plasma cholesterol concentration found in the British Lipid Screening Project. (Mann, J. *et al.* (1984). *British Medical Journal*, **296**, 1702–6.

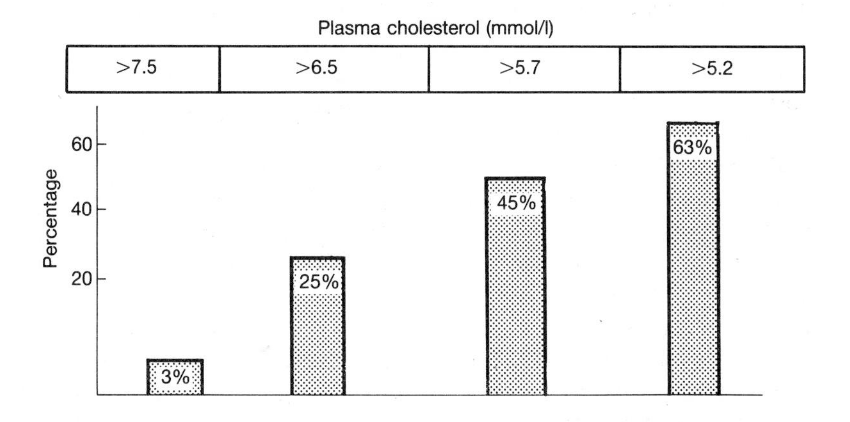

Fig. 2.5. Plasma cholesterol levels in the British population.

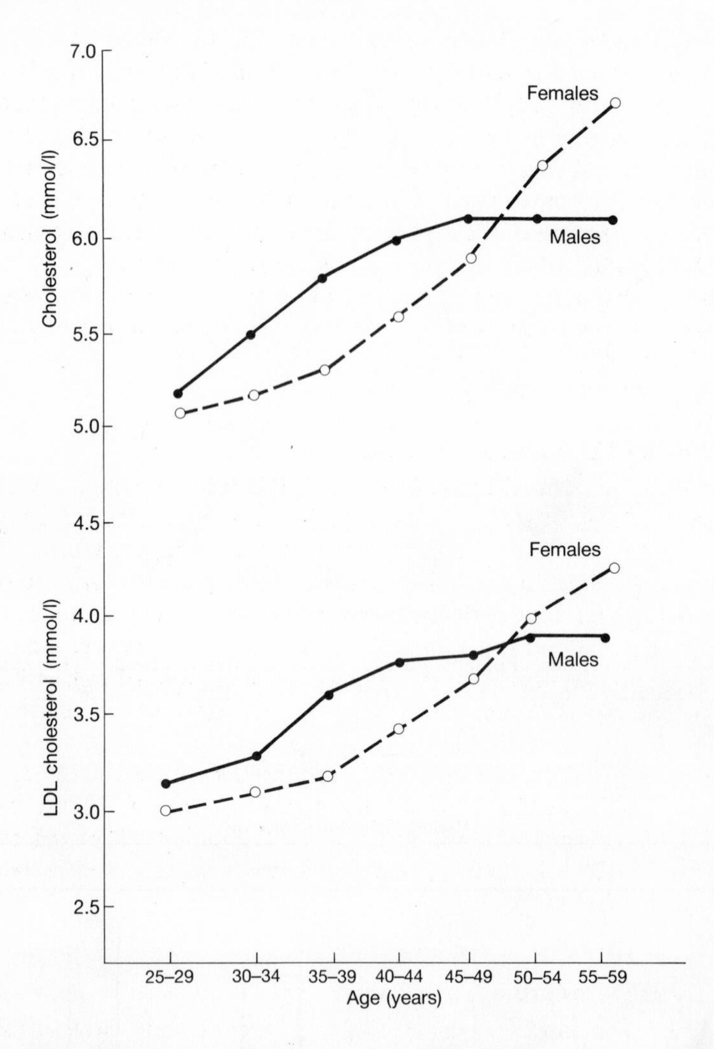

Fig. 2.6. Relationship between total and LDL cholesterol, sex, and age.

The effect of particularly low plasma cholesterol

The inverse relationship between cholesterol and total non-cardiovascular mortality at the lower end of the range (Fig. 2.2) was initially used as an argument against the suggestion that the lowest cholesterol levels were the most satisfactory. There appears to be an increased risk of non-cardiovascular deaths (chiefly from cancer) in those with the lowest plasma cholesterol. Most studies show that this inverse association is confined to deaths in the very early years of follow-up and it is now generally believed that the low cholesterol in some of these individuals was a metabolic consequence of the cancer which was present, but unsuspected, at the time of the initial examination. Further evidence against an unfavourable effect of low plasma cholesterol comes from cross cultural comparisons: countries with a low population mean cholesterol not only have low CHD rates, but also show no evidence of any increased risk of cancer.

Low density lipoprotein cholesterol

The association of total cholesterol with CHD morbidity and mortality appears to derive chiefly, if not entirely, from the LDL cholesterol fraction with which it is very highly correlated. Unfortunately, LDL cholesterol has been measured in relatively few epidemiological studies and there are no clearly defined, internationally accepted, normal or optimal ranges. What is apparent though is that levels greater than 5.0 mmol/l, in adults, are associated with a considerable CHD risk and that most individuals with familial hypercholesterolaemia have levels above this.

The mean LDL cholesterol levels, at different ages, seen in the National Lipid Screening Project are shown in Figure 2.6. In this study the LDL cholesterol was calculated from the measurements of total cholesterol, HDL cholesterol, and triglycerides using the Friedewald equation (see Chapter 5).

High density lipoprotein cholesterol

There has been much interest in HDL as a protective factor against CHD. Women have higher HDL levels than men and a lower CHD risk. The data in Table 2.1, taken from the Framingham Study, show that those with high levels of HDL had an appreciably lower rate of CHD over the 10 years of follow-up than those with lower levels. These figures have led some laboratories to suggest that HDL levels greater than 1 mmol/l are desirable. However, not all prospective epidemiological studies have found HDL to be independently related to CHD. For

Table 2.1. Levels of high density lipoprotein HDL cholesterol and subsequent incidence of ischaemic heart disease in the Framingham study

HDL cholesterol		CHD rate/1000 population
mg/dl	mmol/l	
All levels		77
<25	<0.65	177
25–44	0.65–1.38	103
45–64	1.4–1.64	54
65–74	1.64–1.90	25

example, the British Regional Heart Study found a rather weak association between total HDL and CHD, which was not statistically significant after controlling for the effects of other CHD risk factors. The number of clinical and laboratory studies which have shown that HDL (in particular the HDL_2 sub-fraction) is inversely related to CHD, still mean that it is reasonable to measure HDL when attempting to build up a detailed lipid risk profile. LDL cholesterol is also a clinically relevant value, possibly more so in view of the fact that it may be modified to a greater extent than HDL by simple dietary measures and drug therapy.

Plasma triglycerides

The study of plasma triglycerides is complicated by the fact that levels show a marked variation in response to meals. It is usual to study fasting levels and the fasting plasma triglyceride related to age and sex is shown in Figure 2.7. The reference interval calculated from these data is 0.5–1.8 mmol/l. Markedly raised plasma triglyceride concentrations, to above 15–20 mmol/l, are associated with a risk of acute pancreatitis. High triglyceride levels may also be associated with an increased thrombotic tendency in any artery, and this may occur even in the absence of marked atherosclerosis. Therefore, it has been suggested that one should consider individual assessment for people with modest elevations of

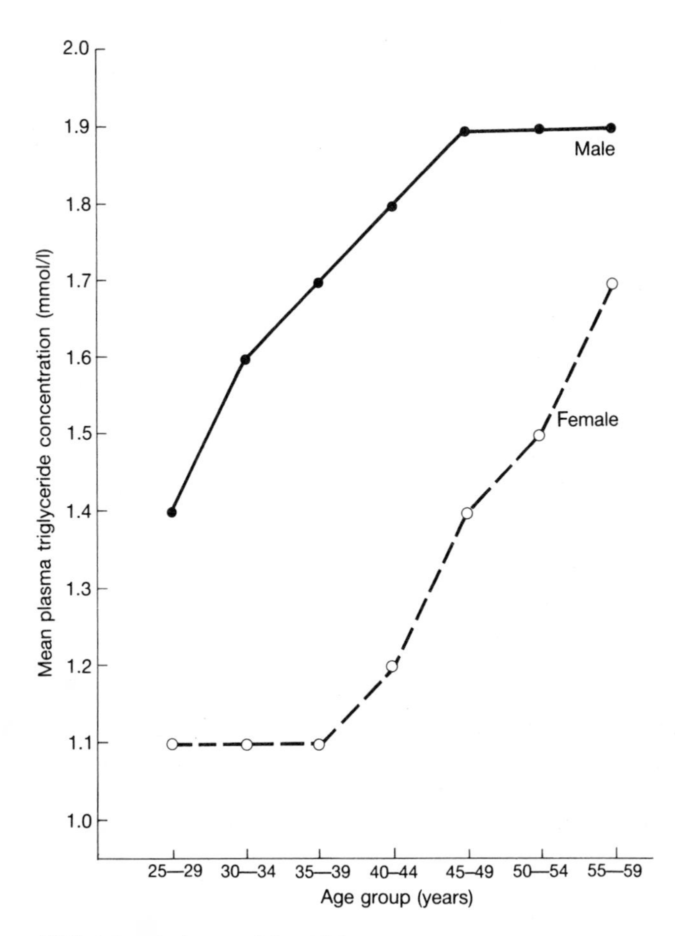

Fig. 2.7. Fasting plasma triglyceride concentrations. (Data from National Lipid Screening Project.)

triglyceride to above 3 mmol/l, and lower if associated with a low HDL level.

The relationship between raised triglycerides and CHD is not clear. Figure 2.8 shows that an increase in CHD risk is principally seen in people with low HDL levels.

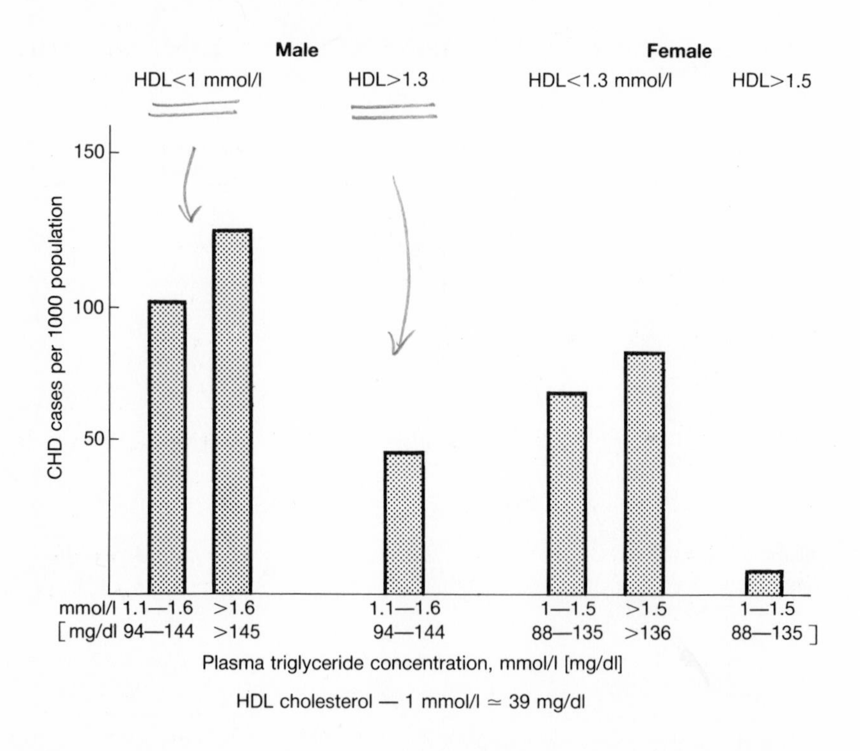

Fig. 2.8. Plasma triglyceride concentration, HDL, and CHD.

Summary

Unlike most other biochemical measurements frequently performed, lipid levels within the 'normal' range are not necessarily optimal. For cholesterol and LDL the 'ideal' range is much lower than the 'normal' range. There appears to be little risk associated with plasma triglyceride levels, which are a little over the upper end of the reference range, but the effect seems to depend on the HDL level. Some of the more important 'normal' ranges and the suggested optimal range are summarized in Table 2.2. Most affluent societies have levels which are appreciably greater than optimal. It seems likely that if optimal levels could be

Table 2.2. Reference and suggested 'healthy' ranges for plasma cholesterol, LDL and HDL cholesterol (mmol/l) and triglycerides (mmol/l) for adults under 60 years

	Reference	Suggested
Total cholesterol	3.5–7.8	<5.5
LDL cholesterol	2.3–6.1	<4.0
HDL cholesterol	0.8–1.7	>1.15
Triglycerides	0.7–1.8	0.7–1.7

Note—The reference interval is the interval calculated from the mean of an apparently healthy population plus and minus two standard deviations.

achieved then the risk of CHD would be appreciably reduced. Evidence for the benefit of cholesterol lowering comes from the clinical trials discussed in Chapter 4.

3 Atherosclerosis: The process and the risk factors

The basic pathology underlying the development of CHD, and other vascular disease, in patients with hyperlipidaemia is nearly always atherosclerosis. Atherosclerosis tends to affect large- and medium-sized arteries such as the aorta and the femoral, coronary, and cerebral arteries, and is common at sites of turbulent blood flow such as arterial bifurcations. As atherosclerosis develops over many years it is not possible to follow the process closely, and there has been great debate as to its pathogenesis.

The current concept of the pathogenesis of atherosclerosis is called the modified response to injury hypothesis. The process is considered to be proliferative, rather than primarily degenerative, and is believed to begin early in life. An accumulation of monocytes and lipid-filled macrophages can be found in the coronary arteries before the age of 10 and this increases during adolescence. The juvenile fatty streak consists of lipid-filled monocytes and macrophages within the intima of the vessel. It occurs, however, in children in all societies, including those in which adult atherosclerosis is rare, and it is controversial whether these lesions progress to fibrous plaques.

Fibrous plaques are white lesions which often protrude into the vessel lumen. The lipid, which is mainly derived from plasma lipoproteins, forms a core, along with necrotic cells, and is covered by a fibrous layer. In advanced lesions, calcium is deposited and thrombus is often present. At least four cell types participate in the formation of an atherosclerotic plaque—endothelial cells, platelets, smooth muscle cells, and tissue macrophages (derived from blood monocytes). Based on studies of diet-induced hypercholesterolaemia in primates, it is believed to be the endothelial cell which is the site where the lesion is initiated, but the actual cause has not been determined. An important event in the process is thought to be an alteration in the functional or structural barrier presented by the endothelial cell lining of the vessel, and it is probable that the properties of these cells are altered as a consequence of hyperlipidaemia. Lipid-filled monocytes can then enter the intima between endothelial cells, and this may further alter the cell properties. Smooth muscle cell proliferation and subsequent migration to the intima is also a focal and crucial process in atherosclerosis.

Experimental damage of arterial endothelium, exposing connective tissue to the blood, causes platelet activation and aggregation on the surface. Platelets then release thromboxanes (primarily thromboxane A), which increase platelet aggregation and lead to the contraction of smooth muscle cells within the arterial wall. Platelet aggregation and adherence to the damaged surface of the atherosclerotic plaque is intimately involved in thrombus formation. Growth factors produced from the endothelial cells, monocytes, and the platelets, can cause proliferation and migration of the smooth muscle cells. *In-vitro* evidence also suggests that LDL stimulates cell proliferation.

The type of atheromatous changes found in vessels of a man dying of premature vascular disease are shown in Figure 3.1 and the possible sequence of events leading to the development of such lesions is shown diagrammatically in Figure 3.2. The result of the process is an increase in the size of the atherosclerotic lesions, which may cause symptoms by reducing blood flow.

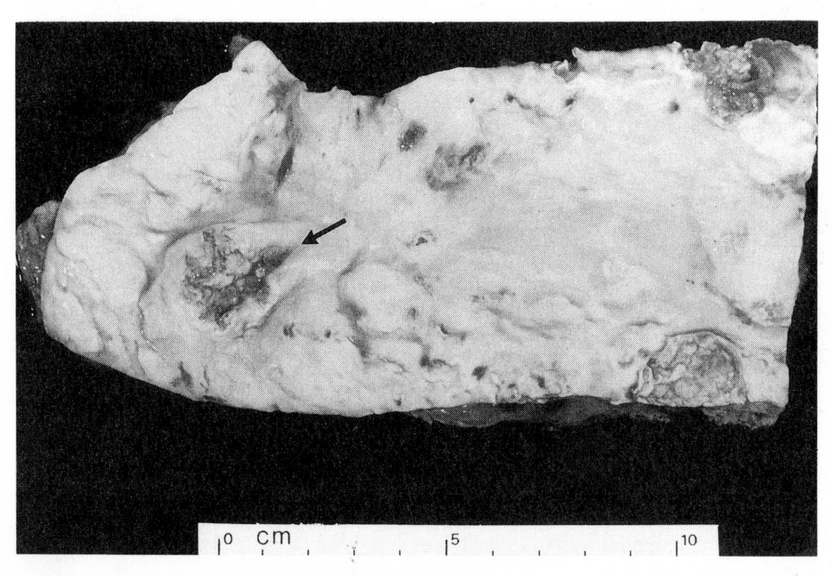

Fig. 3.1. Atheromatous aorta showing fibrous plaques. The lesion indicated has ulcerated.

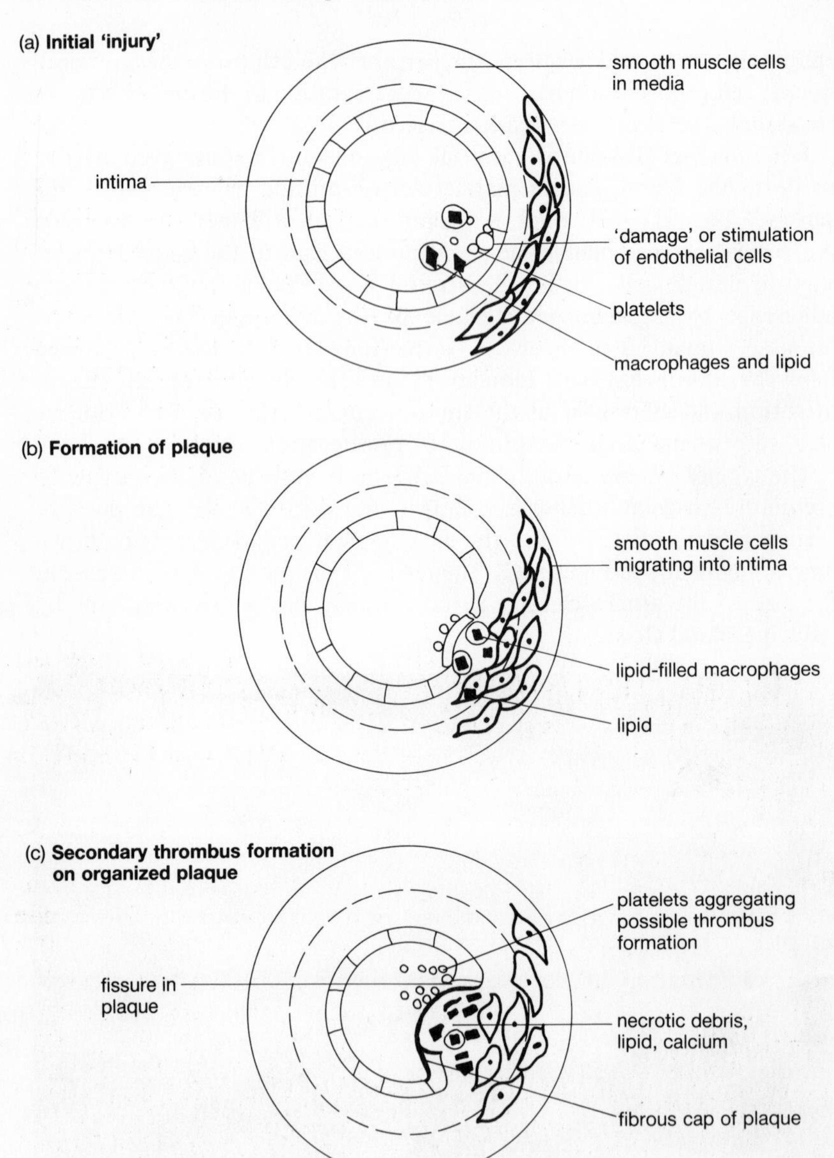

Fig. 3.2. Schematic representation of the possible changes occurring in a vessel to cause atheromatous lesions and potential vessel occlusion.

The causes of myocardial infarction

Acute myocardial infarction nearly always occurs in people who already have significant coronary artery disease. The vulnerability of such individuals to actual infarction depends on a number of different factors. It is often the formation of thrombus in areas of vessels already significantly narrowed by atherosclerosis that causes the acute arterial occlusion resulting in the myocardial or cerebral infarction. This has been shown angiographically to be present in most cases of myocardial infarction and may account for the association between CHD events and plasma concentrations of fibrinogen and some clotting factors. The actual thrombotic incident may be precipitated by fissure of an atherosclerotic plaque which would provide the acute stimulus to platelet aggregation and thrombus formation. It is also possible that less well understood factors contribute, such as coronary spasm.

Risk factors for atherosclerosis and coronary heart disease

Age, sex, cigarette smoking, hypertension, hyperlipidaemia, diabetes, and obesity are the most important factors associated with the development of atherosclerotic disease. Most of the risk factors have a stronger effect in younger people. The relationship of CHD with hypertension and with plasma lipid levels is almost linear.

Age and sex

Atherosclerosis of the coronary and cerebral circulation tends to occur later in life in women, but it increases with age in both sexes. The precise mechanism by which premenopausal women are protected against CHD is not understood, but the rate is much lower in young women than men of a comparable age. CHD frequency increases rapidly after the menopause, but at no age does the rate exceed that of men.

Hyperlipidaemia

The risk of developing atherosclerotic heart disease is directly correlated with plasma cholesterol concentration and up to a third of the weight of atherosclerotic plaques is cholesterol. The mechanism by which cholesterol accumulates in the plaques is not clearly understood, but it appears to be derived from the plasma. This could cause an alteration in the cholesterol to phospholipid ratio in the membranes. Circulating lipoprotein, particularly LDL, can be detected in normal vessel walls in very low concentration. It is possible that if a breach occurs in the endothelium, due to physical or chemical damage, it would allow an influx of LDL into the intima. Lipoproteins could also be absorbed into the intimal connective tissue matrix and be taken up by cells of the arterial wall. High concentrations of LDL or VLDL may thus result in lipid accumulation in the vessel wall. The positive relationship between total and LDL cholesterol and triglyceride, and the inverse association between HDL cholesterol and CHD, were discussed in more detail in Chapter 2.

People with hyperlipidaemia often have another risk factor for CHD. Identification and treatment of these factors is most important. It appears that raised plasma lipids enhance the damaging effects of other parameters. For example, in Japan hypertension is quite common, but CHD is relatively rare in those who eat the basic Japanese diet which is low in saturated fat. However, if they move to an area where they consume a diet in saturated fat then their risk of developing CHD increases.

Cigarette smoking

Smoking is strongly related to cardiovascular disease in countries where CHD rates are high (Fig. 3.3). Large prospective studies have shown that men who smoke more than 20 cigarettes a day have three to four times the risk of dying from CHD than those who do not smoke. Furthermore, the relationship between smoking and CHD is linear. Smokers also have an increased risk of stroke and intermittent claudication. Sudden cardiac death is quite frequently the first clinical manifestation of CHD, and is three times more common than in non-smokers. Atherosclerotic changes found at post-mortem correlate with smoking habits, and the association applies to changes in the aorta, the large coronary arteries, and the small intramyocardial arteries. Heavy cigarette smokers with high carboxyhaemoglobin levels have a particularly high risk of vascular disease, and may develop symptoms at an earlier stage of the disease because of the slightly lowered blood oxygen content. Cigarette smokers who undergo coronary artery bypass have a higher perioperative mortality than non-smokers.

Figure 3.3 shows that the association with cigarette smoking is much less striking amongst Southern European men. Smoking more than

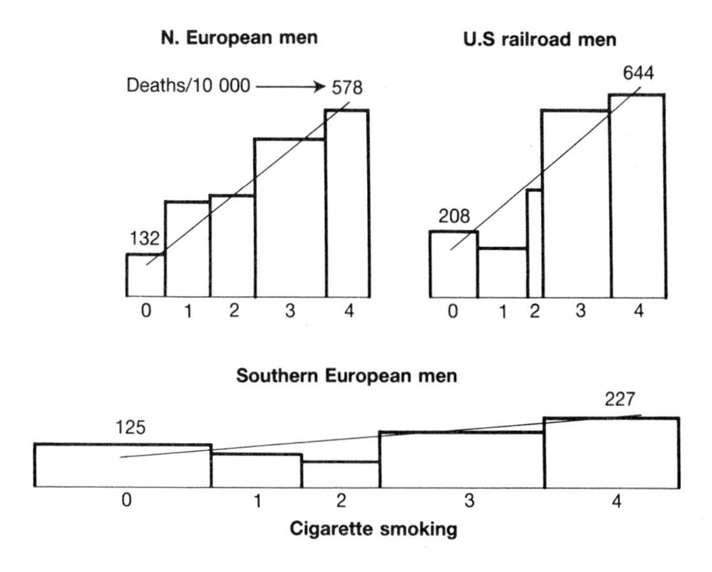

Fig. 3.3. Relationship of smoking and CHD deaths in different countries. (From Keys, A. Seven countries. A multivariate analysis of death and coronary heart disease. Harvard University Press, 1980.)

20 cigarettes per day among this group is associated with only a two-fold increase of fatal CHD compared with non-smokers. This group have a lower saturated fat intake and cholesterol level suggesting that the higher plasma lipids in the Northern Europeans and Americans enhances the damaging effect of other risk factors. Again, in Japan, where both cigarette smoking and hypertension are common, but cholesterol levels are low, CHD occurs relatively infrequently.

Hypertension

A near linear association is found in high-risk populations between systolic and diastolic blood-pressure and CHD. As with cholesterol it is difficult to identify a cut-off point associated with a particularly high risk. The Framingham Study showed that a systolic blood-pressure greater than 160 mmHg or a diastolic greater than 95 mmHg carries a two- to three-fold increased risk of CHD, and an even greater risk of cerebral vascular disease. Treatment of patients with moderate hypertension seems to reduce the overall mortality, but this is mainly by reducing the incidence of strokes. The interactive relationship between risk factors—systolic blood-pressure and smoking—is shown in Figure 3.4.

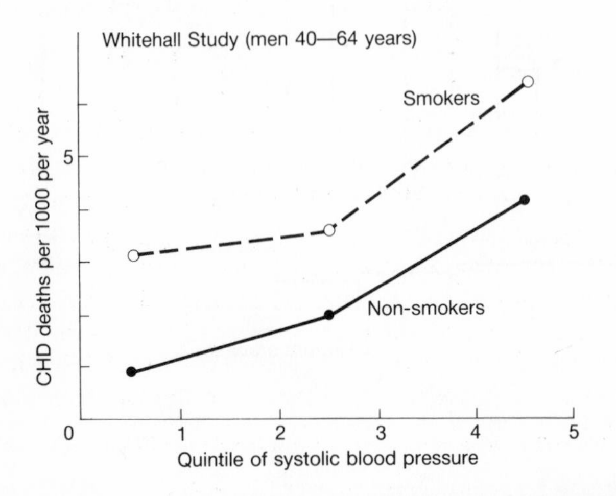

Fig. 3.4. Smoking and systolic blood-pressure as risk factors for CHD. From the Whitehall study of British civil servants (Reid *et al.* (1976). *Lancet* **ii**, 979–84.)

Diabetes or glucose intolerance

Individuals with diabetes or impaired glucose tolerance (as defined in Chapter 6) are at substantially increased risk of CHD, although the relationship with a single glucose measurement is less convincing. Women are particularly susceptible to the effects of diabetes. In diabetic women the usual protective effect of the female sex is lost. The presence of a high insulin level is also associated with the subsequent development of CHD. The precise mechanism by which diabetes increases the risk of CHD has not been established. It is not clear whether basement membrane abnormalities contribute.

Obesity

Significant obesity confers an increased risk of CHD, but it is unclear to what extent this is an independent risk factor as obese people often have raised blood-pressure, plasma glucose and plasma lipids. Obesity is, however, common in many Western countries and whether its influence is secondary or not, obesity is a useful way of identifying those at increased CHD risk. A total of 34 per cent of men and 24 per cent of women in the U.K. are overweight, with a body mass index (weight/height^2) greater than 25, and 7 per cent are seriously overweight, with a body mass index greater than 30.

Exercise

Exercise is difficult to evaluate as an independent variable since those taking regular exercise are self-selected and exercise can affect other important variables. Nevertheless studies have shown that those involved in regular, heavy, physical work have a reduced risk of sudden death. Those who engage in vigorous exercise in their leisure time also have a lower incidence of myocardial infarction than inactive individuals. In addition, people who take regular exercise appear to have a slightly lower risk of developing hypertension, and exercise can be of value in those attempting to lose weight. Regular vigorous exercise has been shown to increase HDL levels slightly.

Ethnic origin and psychosocial factors

People of Asian origin living in Britain have a higher CHD mortality than Caucasians. The mortality is higher than those living in their country of origin, perhaps reflecting the effects of genetic predisposition plus the acquisition of other factors while living in Britain. People from the Caribbean also have a high mortality from stroke and hypertensive disease.

Mortality from CHD is higher overall in unskilled workers (social classes IV and V) than in the professional classes and the gap has widened in the last decade (Fig. 3.5). This is partly accounted for by a different frequency of other risk factors in these groups.

There has been debate about the role of 'stress' in the development of CHD and about what constitutes stress as precise measures have proved elusive. In the Whitehall Study of British civil servants there was a step-wise inverse relationship between employment grade and CHD mortality. Those in the lower grades did tend to smoke more, had a higher mean blood-pressure, less leisure time physical activity, and a higher prevalence of diabetes and obesity, but it appeared that this did not account for the entire difference in CHD. Job characteristics such as

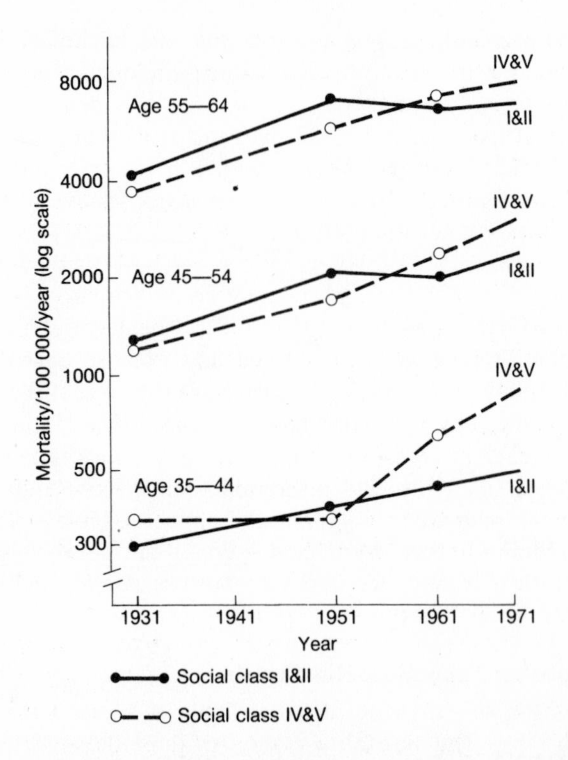

Fig. 3.5. Relationship between social class and CHD. (From Marmot *et al.* (1978). *British Medical Journal*, **ii**, 1109–12.)

'lack of control' over the work and poor social support may be important. Also relevant in the current economic climate is the fact that longitudinal studies show the unemployed to have a 20 per cent higher mortality rate from CHD than the employed.

Personality factors

Some studies have indicated that people with a type-A behaviour (aggressive, ambitious, and restless, with an increased sense of time-urgency) have a greater risk of CHD and sudden death than those with a more passive type-B behaviour. This concept was devised in white Caucasian males and is difficult to evaluate. The relationship in this group may be partly related to other risk factors in some of the men with type-A behaviour. Of note is the apparent contradiction between this theory and the findings of studies such as the Whitehall Study, where type-A behaviour was commoner in those of higher grades yet they had a lower overall CHD mortality.

Clotting factors

The effects of the clotting factors were extensively studied in the Northwick Park Prospective Study. Factor VII coagulant activity and fibrinogen appear to be strongly related to the subsequent development of CHD and the predictive effect may be as strong as that of plasma cholesterol. Genetic factors may be important, but it is of interest that fibrinogen levels are increased by cigarette smoking and factor VIIc levels are related to the intake of saturated fat.

Genetic factors

A tendency for CHD to run in families has been recognized for a long time. It was initially assumed that inherited forms of hyperlipidaemia and inherited factors in diabetes, and possibly hypertension, accounted for this finding. Several studies have now shown that the familial aggregation of CHD is not exclusively mediated by familial resemblance in plasma cholesterol and other classical risk factors. The risk related to a strong family history of CHD was demonstrated in the Framingham Study where analysis showed that the incidence of myocardial infarction in brothers was significantly related after the effects of cholesterol, blood-pressure, and smoking had been controlled. It is possible that a tendency to hypercoagulability may be present in some of the families with an increased incidence of premature CHD and no obvious risk factors, but it is likely that there are a number of other important genetic factors. Recent advances in techniques in molecular biology are allowing investigation of some genetic factors which may be related to the development of premature CHD. Such techniques are likely to increase our understanding of the high degree of heritability of the disease.

Summary

Atherosclerosis is usually the basic pathology underlying the development of symptomatic CHD. There are a number of factors which appear to accelerate the development of atherosclerosis. Elevated plasma cholesterol is of major importance and it also appears to enhance the adverse effects of other risk factors.

4 Does lipid lowering reduce coronary heart disease risk?

Populations with low mean cholesterol levels have low rates of CHD and within high-risk populations there is a gradient of CHD risk with increasing levels of cholestrol. Studies on migrants show that the CHD risk of populations can be modified by changes in diet, blood cholesterol, and other environmental factors. When Japanese, who traditionally have a low risk of CHD, migrate to Hawaii or the USA they soon tend to acquire rates which approximate to those in the host country. On the other hand, Finns who live in Sweden, where CHD rates are relatively low, have appreciably lower rates than when living in their native Finland where CHD rates are among the highest in the world. What needs to be known, however, is the clinical benefit to individuals of cholesterol lowering by dietary manipulation or drugs. This chapter summarizes the clinical trials which have been performed to establish whether cholesterol lowering over a relatively short period of time can achieve a reduction in morbidity and mortality from CHD, or cause regression of atherosclerosis.

Effect of lipid-lowering treatment on angiographic change

A double-blind randomized study in the USA assessed the regression of atherosclerosis by coronary angiograms. Patients with hypercholesterolaemia associated with raised levels of LDL received either dietary advice and cholestyramine (24 g/day), or dietary advice and placebo. Coronary angiograms were performed before and after five years of treatment. Progression of the lesions occurred in 49 per cent of the placebo group, but in only 32 per cent of those treated with cholestyramine, who achieved a 26 per cent reduction in LDL cholesterol compared with the placebo group. Of the various lipid measurements made, the ratio of HDL to LDL cholesterol was the best predictor of angiographic change, the lowest ratio being associated with the greatest risk of progression of atherosclerosis. Two other studies have been carried out in Finland (where clofibrate, nicotinic acid, or, if necessary, both drugs were used to achieve lipid lowering) and the Netherlands (where a strict vegetarian diet was used to lower lipids). These investigations were less carefully controlled than the study in the USA but once again suggested that cholesterol lowering could halt the progression of atheroma. A British study also suggested that the development of atherosclerosis in other vessels, in this case the femoral artery, could be slowed. These findings are encouraging. Computer-assisted assessment of coronary angiograms and more sophisticated non-invasive techniques for quantifying atherosclerosis may give more definitive results.

Trials of diet to control hypercholesterolaemia

Early diet studies

Most studies investigating dietary manipulations known to lower blood lipids have been too small to produce meaningful conclusions in terms of morbidity and mortality, as the confidence limits of the results are very wide. Only one of these early studies warrants individual mention. In the Los Angeles Veterans Administrative Study, 846 male volunteers (aged 55–89) received either a 'control' diet (40 per cent energy from fat, mostly saturated, typical of the North American diets) or an 'experimental' diet (with half as much cholesterol and predominantly polyunsaturated vegetable oils replacing approximately two-thirds of the animal fat). As a result of skilled food technology the study was conducted under double-blind conditions. Follow-up was eight years.

It was found that:

- cholesterol levels in the experimental group were 13 per cent lower
- deaths due solely to atherosclerotic events were appreciably reduced as compared with the controls (see Table 4.1)
- the beneficial effect of the cholesterol-lowering diet was most marked in those with initial high plasma cholesterol levels.

Deaths due to conditions other than CHD and from uncertain causes occurred more frequently in the experimental group, though no single cause predominated. The increase in non-cardiovascular mortality in the experimental group raised for the first time the suggestion that cholesterol lowering might be harmful in some respects despite the reduction in CHD. This was discussed in Chapter 2.

Table 4.1. Summary tabulation of deaths by category in the Los Angeles Veterans Administration Study

Category	Number of cases	
	Control	Experimental
Due to acute atherosclerotic event (sole cause)	60	39
Mixed causes, including acute atherosclerotic event	10	9
Due to atherosclerotic complication without acute event	1	2
Mixed causes, including atherosclerotic complication with acute event	10	7
Other causes	71	85
Uncertain causes	25	32
Total	177	174

(From Dayton, S. *et al.* (1969). *Circulation*, **39**, suppl. 11, 1–63.)

Multiple factor intervention studies

European Collaborative Trial

A multi-centre trial introduced health education into selected factories in four countries. Individuals in different factories were paired; one of each pair received education and the other acted as a control. The health education included advice on diet (aimed at cholesterol lowering and reduction of obesity), emphasized the importance of not smoking and of increased physical activity, and gave information on hypertension. In addition, men with a mean systolic blood-pressure above 160 mmHg were started on anti-hypertensive drug therapy.

It was found that:

- only small net reductions in risk factors were achieved
- the reduction in overall CHD mortality was only 7.4 per cent

Support that this represented more than just a chance improvement came from the results from individual centres. The U.K. had the least risk-factor reduction and showed no evidence of a fall in CHD incidence. Belgium and Italy produced the greatest reduction in risk factors and greater changes in CHD incidence: for example, in Belgium, there was a highly significant 24 per cent reduction in CHD.

Oslo Trial

In the Oslo Trial, men at high risk of CHD (as a result of smoking or having a cholesterol level in the range of 7.5–9.8 mmol/l) were divided into two groups; half received intensive dietary education and advice to stop smoking, the other half served as a control group.

It was found that:

- a reduction (31 *vs.* 57 per 1000 over a five-year period) in total coronary events occurred (see Fig. 4.1) in association with a 13 per cent fall in cholesterol and a 65 per cent reduction in tobacco consumption

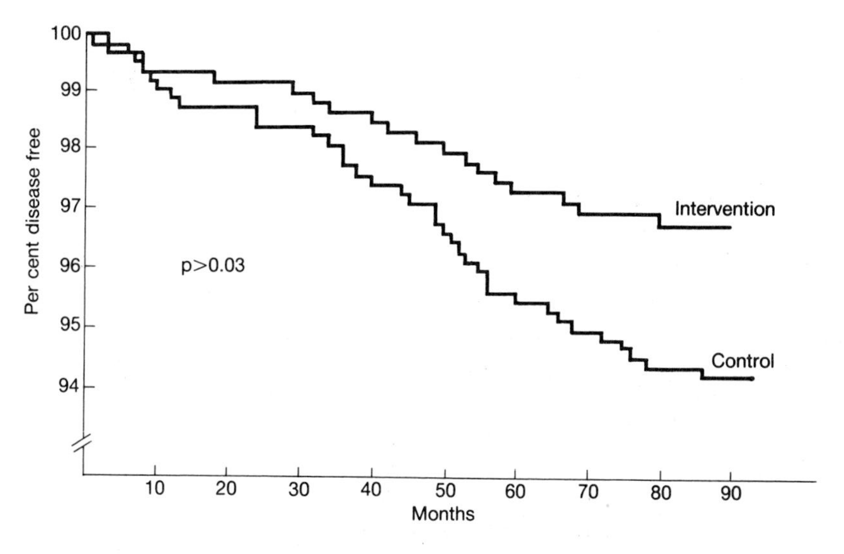

Fig. 4.1. Life table analysis of CHD (fatal and non-fatal myocardial infarction and sudden death) in intervention and control groups. The Oslo Study. (From Hjermann, I. *et al.* (1981). *Lancet*, **ii**, 1303–10.)

- there was a significant improvement in total mortality in the treated groups, and no significant differences between the two groups for non-cardiac causes of death

Detailed statistical analysis suggests that approximately 60 per cent of the CHD reduction can be attributed to serum cholesterol change and 25 per cent of smoking reduction.

Multiple risk factor intervention trial

In the light of the relatively encouraging results from Europe the findings of the American Multiple Risk Factor Intervention Trial (MRFIT) were rather disappointing. Men at high risk were chosen on the basis of plasma cholesterol level and smoking (similar to the Oslo study) and of high blood-pressure. Intervention against these three risk factors for six years in the 'special intervention' group was compared with the control group randomized to 'usual care'. The results are shown below.

	Intervention group	Control group
Smoking	Reduction of 50 per cent	Reduction of 29 per cent
Diastolic B.P.	Fall of 10.5 mmHg	Fall of 7.3 mmHg
Serum cholesterol	Fall of 5 per cent	Fall of 3 per cent

The changes in the control group were presumed to result from the widespread coronary prevention education in the USA.

Clearly, a trial which achieves a reduction in plasma cholesterol in the intervention group only 2 per cent greater than the control group cannot be used to determine whether a reduction in plasma cholesterol will lead to a reduction in CHD incidence. Furthermore, over the study period, CHD mortality in the USA declined by 25 per cent. As a result of this, and of the risk-factor reduction in controls as well as the intervention group, both groups had lower than predicted mortality and there were no significant differences between the groups.

When one considers only hypercholesterolaemic men or men who smoked (i.e. individuals similar to those participating in the Oslo trial), however, the improvement noted in the special intervention group was comparable with that observed in Oslo. The higher mortality amongst hypertensive men, especially those with electrocardiograph abnormalities, has led to the suggestion that the drugs used for treating hypertension may have adverse effects.

The MRFIT underlines the near impossibility of achieving an appropriate control group now many people are aware of coronary risk factors and their consequences. Further large dietary trials (either single-factor or as part of multi-factorial intervention) are unlikely to be carried out.

Trials of lipid-lowering drugs

Four major trials using drugs to reduce lipid levels have been carried out.

Coronary drug project

A total of 8000 men with a history of myocardial infarction were randomized to five active treatments or placebo.

It was found that three of the therapies (conjugated oestrogens 2.5 and 5.0 mg daily, dextrothyroxine 6.0 mg/day) were associated with an excess mortality in comparison with placebo. Clofibrate (1.8 g/day) was associated with a non-significant reduction of CHD and niacin (3 g/day) with a significant reduction in non-fatal myocardial infarction which has persisted during a 15-year follow-up.

WHO co-operative trial of clofibrate

A total of 15 745 healthy men aged 30–59 were allocated to three groups on the basis of their cholesterol levels. Half of those in the upper third of the cholesterol distribution were randomly assigned to clofibrate treatment (Group 1) and the other half to indistinguishable olive oil capsules (Group 2). A second control group (Group 3), chosen randomly from the lowest third of the cholesterol distribution, was also given olive oil. The trial was carried out under double-blind conditions.

It was found that:

- during the five years of active therapy a 9 per cent reduction of cholesterol was achieved on clofibrate
- there was a significant reduction of major CHD in Group 1 compared with Group 2
- the difference between the two groups was due to cholesterol lowering

- Men with the highest initial cholesterol, the greatest reduction during the trial, and those with other CHD risk factors, had the greatest reduction in non-fatal CHD.

There was, however, a significant increase in mortality in the clofibrate group. This was due to a variety of causes, with no particular disease (except for gallstones) predominating, and there was no relation between excess mortality and cholesterol reduction. The excess mortality in the clofibrate group did not continue after discontinuation of the drug.

The initial hypothesis behind the trial was that reduction of plasma cholesterol would reduce the incidence of CHD; this was confirmed. Clofibrate was merely the chosen method of reduction. The excess of deaths in the 'treated' group has diverted attention from this result. Furthermore, the data concerning total mortality are difficult to interpret since a substantial number of deaths were not reported in the publications. As discussed elsewhere, there is no evidence for an adverse effect of cholesterol lowering *per se*.

Lipid research clinics coronary primary prevention trial

This multi-centre, randomized, double-blind study tested the efficacy of cholesterol lowering in reducing the risk of CHD in 3806 asymptomatic men with primary hypercholesterolaemia which had not responded to simple dietary management. The treatment group received the bile acid sequestrant cholestyramine and the control group received a placebo, for an average of 7.4 years.

The cholestyramine group experienced a mean reduction in total plasma and LDL cholesterol of 13.4 per cent, and 20.3 per cent respectively.

In comparison with the placebo group there was:

- a 24 per cent reduction in fatal CHD,
- a 19 per cent reduction in non-fatal myocardial infarction
- an appreciably reduced incidence rate for new positive exercise tests, angina, and coronary bypass surgery (Fig. 4.2).

The reduction of CHD incidence in the cholestyramine group appeared to be mediated chiefly by the fall in total and LDL cholesterol.

Accidents and violence occurred more frequently (though not

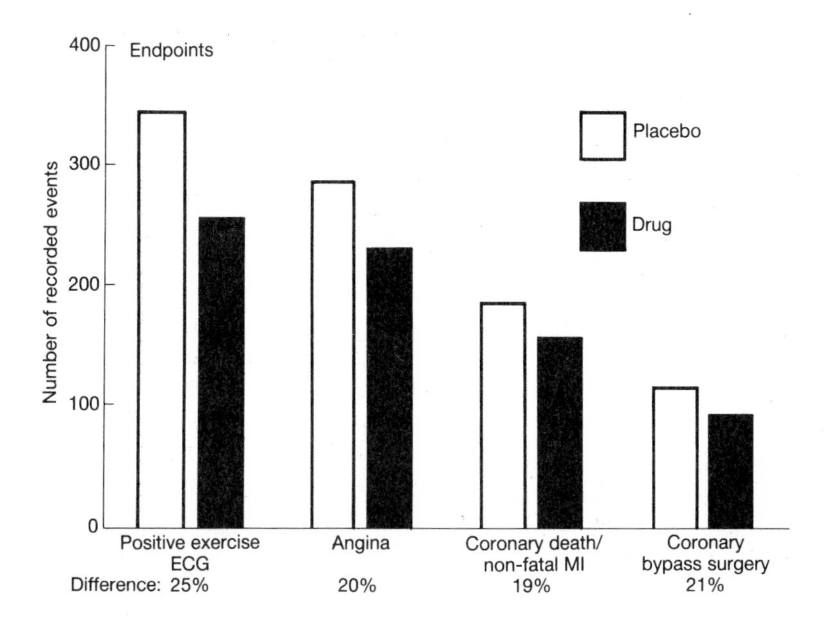

Fig. 4.2. Results of the Lipid Clinics Primary Prevention Trial.

significantly so) in the cholestyramine group, but deaths from all other causes were similar in the two groups.

The Helsinki Heart Study

Inclusion in this study was based on the level of the non-HDL cholesterol (LDL + VLDL cholesterol): 4081 men with a non-HDL cholesterol >5.2 mmol/l were randomized to receive gemfibrozil 600 mg b.d or placebo for a five-year period. Gemfibrozil gave a mean reduction of 8 per cent for plasma cholesterol, 9 per cent for LDL cholesterol, and 40 per cent for plasma triglycerides. HDL cholesterol increased by about 11 per cent.

A difference in the incidence of cardiac disease between the two groups was seen after two years. The cumulative rate of cardiac end-points at five years was 27 per 1000 in the gemfibrozil treated group and 44 per

1000 in the placebo group. The reduction in CHD in the treated group was 34 per cent.

Overall perspective of the trials

A helpful perspective of the lipid-lowering trials has been provided by Peto *et al.* (1987). They pointed out that in all the studies (especially the smaller ones) the final results are subject to a considerable margin of error. Consequently, they have fitted confidence limits to the differences in CHD rates between the control and experimental groups and pointed out the importance of considering the results of the various trials together. When this is done, CHD reduction is associated with the extent of the fall in cholesterol. After two years a 5–9 per cent fall in cholesterol is associated with an average 8 per cent reduction in CHD; when the decrease is 10–15 per cent, a 19 per cent CHD reduction occurs. Similarly when considering the short trials, a 10 per cent difference in cholesterol between the experimental and control groups would produce an average 11 per cent CHD reduction, whereas in the longer trials (three years or more) the same cholesterol difference is associated with a 21 per cent reduction in CHD. Figure 4.3 shows the drop in risk of CHD in the various trials in relation to achieved cholesterol lowering (presented as strength of intervention calculated by multiplying the cholesterol difference between experimental and control groups by years duration). The association between the strength of intervention and reduction in CHD risk is impressive, providing strong confirmatory evidence for the importance of cholesterol in the aetiology of CHD. It suggests that cholesterol lowering even in middle-aged individuals can be of benefit in reducing coronary events.

The data are particularly encouraging when considering patients with appreciably raised cholesterol and no pre-existing CHD. The Oslo Trial, the WHO Co-operative Clofibrate Trial, the Lipid Research Clinics Trial of cholestyramine, a Swedish study by Rosenhamer and Carlson (of clofibrate and nicotinic acid), and the Helsinki Heart Study all show appreciable reduction of CHD incidence in association with cholesterol lowering. Of the trials which included those with high cholesterol levels, the MRFIT study was the only one not associated with an overall reduction in coronary events, perhaps because there was a relatively small mean decrease in cholesterol and virtually no difference between the intervention and control groups.

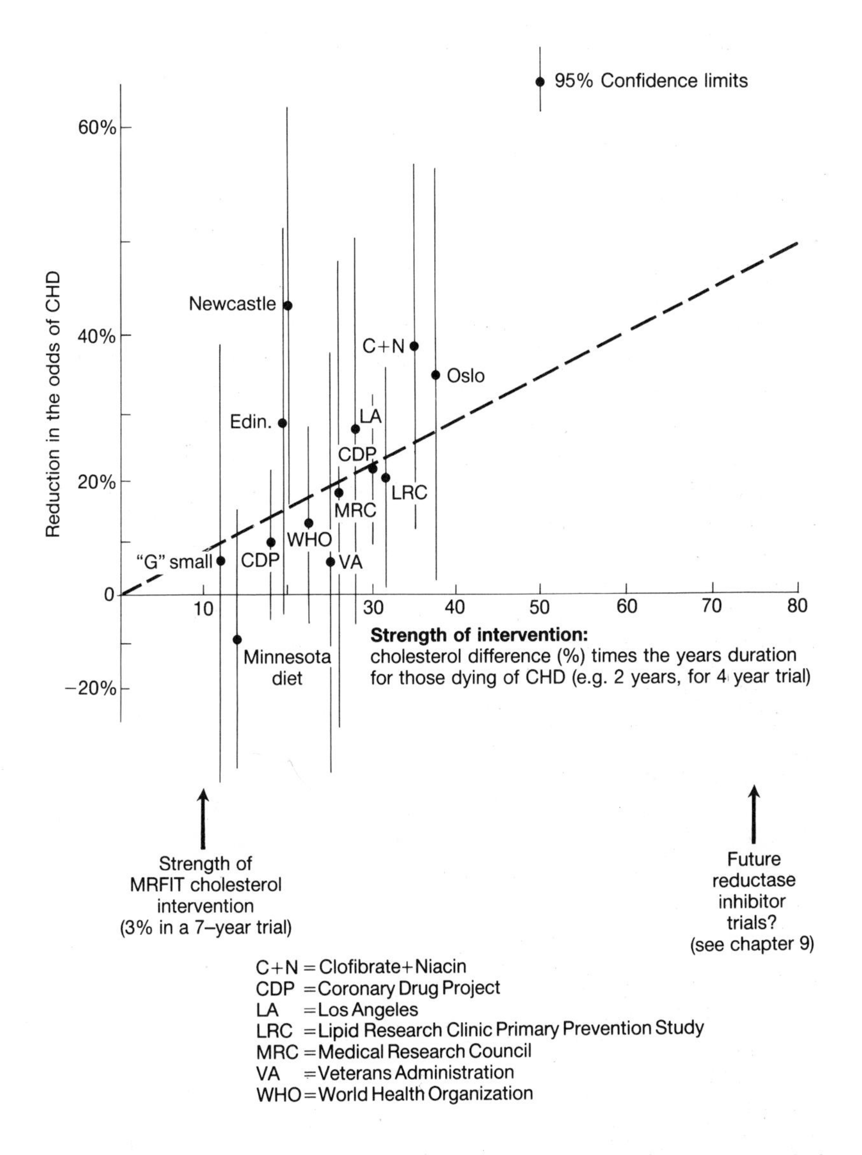

Fig. 4.3. CHD reduction *versus* strength of intervention in the unconfounded randomized trials. (Peto, R. *et al.*, reproduced with permission.)

Table 4.2. Overall estimates of effect on total CHD of a 'standard' 10 per cent cholesterol reduction in a four-year trial, using various lipid-lowering methods.

	No. of trials	Estimated CHD reduction
Clofibrate	4	16% ± 5%
Niacin	2	14% ± 5%
Clofibrate & niacin	1	24% ± 9%
Bile acid sequestrants	5	15% ± 7%
Diet	8	13% ± 6%
All unconfounded randomized trials	20	16% ± 3%

The beneficial effect of cholesterol lowering appears to occur regardless of the means by which lowering is achieved. Table 4.2 shows the aggregated data for the 20 unconfounded randomized trials. The overall 16 per cent reduction in CHD is associated with a 95 per cent confidence interval of 11–20 per cent. Some writers have drawn attention to the fact that several studies have shown an increase in some non-cardiac causes of mortality and that, as a result, total mortality has not been reduced. This has not been fully elucidated. Peto has pointed out that apart from the well known risk of gallstones with clofibrate and increased total mortality in the WHO Co-operative Study of clofibrate, the increased mortality from any individual non-CHD cause has been trivial when compared with the reduction in incidence and mortality from CHD. It has been non-specific (no single cause of death being predominant or consistent) and not statistically significant. Mortality from non-cardiac causes is not related to the extent of cholesterol lowering and has not been in keeping with the epidemiological evidence. The benefit of cholesterol lowering on CHD on the other hand is specific, highly statistically significant, proportional to the cholesterol lowering achieved, and in keeping with the epidemiological evidence.

Summary

Differences in cholesterol levels and saturated fat intake correlate with CHD frequency in different countries and cultures. Cholesterol lowering by diet and by drug therapy reduces CHD.

Section II

Hyperlipidaemia: diagnosis and management

5 Hyperlipidaemia: diagnosis and assessment

The presence of raised plasma lipids alone seldom causes any symptoms or physical signs until the secondary pathologies of atherosclerosis or pancreatitis occur. Yet hyperlipidaemia, which poses a long-term threat to health, is very common. As discussed in other sections, a reduction in plasma cholesterol appears to be effective in reducing the incidence of CHD, both in individuals and in populations.

The data in Chapter 2 shows that it is impossible to identify thresholds for the unequivocal definition of hyperlipidaemia as there is a continuous gradient of risk associated with increasing plasma cholesterol levels (Fig. 5.1). For this reason various organizations have identified 'action limits'. It is not really surprising that there is some variation in the recommendations of different bodies. To some extent most recommendations are based on the practical consideration that it would be impossible to deal individually with everyone at some degree of risk. The levels below are based on those suggested by the European Atherosclerosis Society.

Cholesterol level (mmol/l)			
Statistical reference range	Desirable	At some CHD risk	Identify for individual care
3.9–7.8	<5.2	5.2–6.5	>6.5

The authors wish to emphasize that these levels are suggested in the realization that those identified for individual care will need attention and follow-up, even though most will be symptom free. If universal screening were to be introduced then about a quarter of the adult population would currently be in this category, which is an awesome prospect. This would include those with severe inherited forms of hyper-

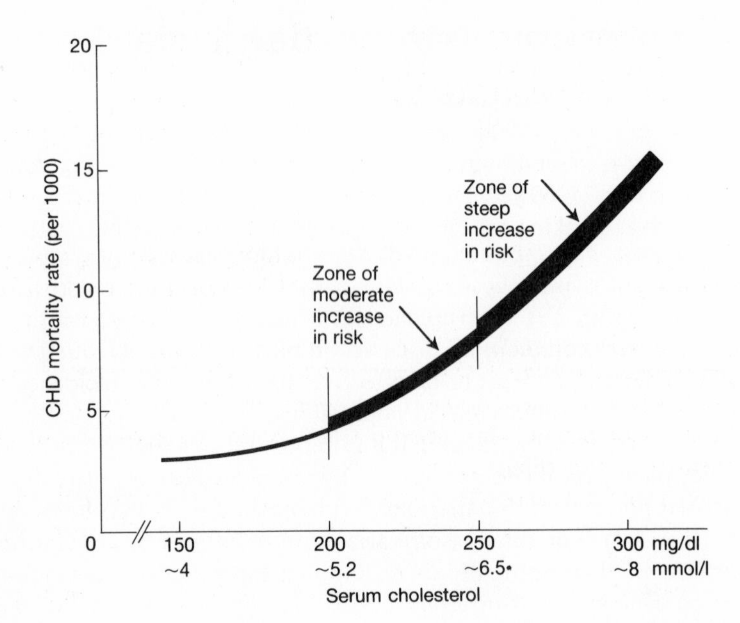

Fig. 5.1. CHD mortality rate in six years (MRFIT (1982), 356 222 men, age 35–57).

lipidaemia who have an exceptionally high risk of CHD, but very many of the group would have high lipid levels because of a high-fat diet, obesity, and/or poorly diagnosed genetic factors. The successful implementation of a population dietary strategy (Chapter 11) would greatly reduce the number with hyperlipidaemia and many others identified would respond to more specific dietary advice and not require drug treatment.

Much less information is available concerning desirable and 'at risk' plasma triglyceride levels. The European Atherosclerosis Society has suggested individual consideration of levels above 2.3 mmol/l. However, for most practical purposes we suggest that in the presence of desirable levels of cholesterol, especially if LDL and HDL levels are satisfactory, that, at present, people should only be identified for individual follow-up if triglycerides are greater than 3 mmol/l. We currently suggest that the initial diagnosis of hyperlipidaemia is based on a total cholesterol concentration greater than 6.5 mmol/l in adults and/or a fasting triglyceride concentration greater than 3 mmol/l.

It is important that ways are found to identify individuals with hyperlipidaemia, preferably before any symptoms of CHD develop. The most direct method is to set up a specific service to screen the entire population in the age group 25–60 years, in conjunction with screening for other risk factors such as smoking and hypertension. In the absence of this comprehensive approach, medical staff should be aware of a number of circumstances when it is particularly important to measure blood lipids.

These circumstances include:

(1) people with proven CHD or other manifestations of atherosclerosis, especially when these develop at a young age (less than 55);
(2) relatives of people who develop CHD below the age of 55 or are themselves hyperlipidaemic;
(3) people with a disease which may secondarily raise blood lipids, such as diabetes, hypothyroidism, nephrotic syndrome, and severe obesity (BMI $\geq$ 30);
(4) people with stigmata of severe hyperlipidaemia such as xanthomas and early corneal arcus.
(5) people with pancreatitis;
(6) people with another CHD risk factor such as hypertension.

If hyperlipidaemia is suspected, a fasting blood sample should be taken and cholesterol and triglyceride levels measured. The results are then interpreted in the light of the patient's medical history, drug history, family history, diet, and age. The patient should also have a physical examination, particularly looking for signs of lipid deposition—notably xanthelasma, corneal arcus, tendon xanthomas, and eruptive xanthomas.

This assessment process can be understood as dividing those with raised lipids into four groups:

1. A large number of people with hyperlipidaemia; as many as a third of the population in some countries, notably the UK. In this group the hyperlipidaemia is the result of a high fat diet in susceptible individuals, which is sometimes accentuated by obesity or a mild degree of inefficiency in lipoprotein clearance. Most of these people will respond readily to a reduction in dietary fat intake to 'sensible' levels (see COMA recommendations). Dietary management is thus the mainstay of their treatment. Drug therapy has no place in their management.

2. People with an underlying illness, either known or covert, which provokes an increase in plasma lipid levels, which may be mild or sometimes very severe. The cause of this secondary hyperlipidaemia must be identified (Chapter 6). These patients sometimes respond to dietary management, but usually the causative illness must be effectively treated before the levels of the blood lipids fall. Occasionally the illness cannot be treated and a specific lipid-lowering diet and drugs have to be considered.

3. A group of people, maybe 1 per cent of the population, with a severe primary disorder of lipid metabolism caused by a major genetic defect (Chapter 7). The hyperlipidaemia is usually severe and the response to simple dietary change generally inadequate. They often have a very strong family history of early CHD and death. They respond to strict and extensive dietary change combined with drug treatment, and the approach should be aggressive once the diagnosis is made;

4. People with primary hyperlipidaemia which does not respond to dietary management and lifestyle changes, but who do not have any obvious family history, or an obviously identifiable biochemical defect. At present they are often said to have 'common hyperlipidaemia', and no doubt future research will delineate a number of different aetiologies in this group.

Figure 5.2 gives a simple outline of the suggested protocol for the assessment and diagnosis of a patient with raised blood lipids. Chapters 6 and 7 discuss the presentation and management of the secondary and primary hyperlipidaemias.

Secondary hyperlipidaemia (Chapter 6) must be excluded at an early stage and appropriate treatment given. Dietary modification (Chapter 8) should be recommended to all people with primary hyperlipidaemia. Those who do not respond to simple dietary advice require further investigation in order to establish a more precise diagnosis (Chapter 7) and to decide upon appropriate drug therapy (Chapter 9).

Laboratory investigation of hyperlipidaemia

Measurement of plasma cholesterol and triglyceride concentrations

Points to consider:

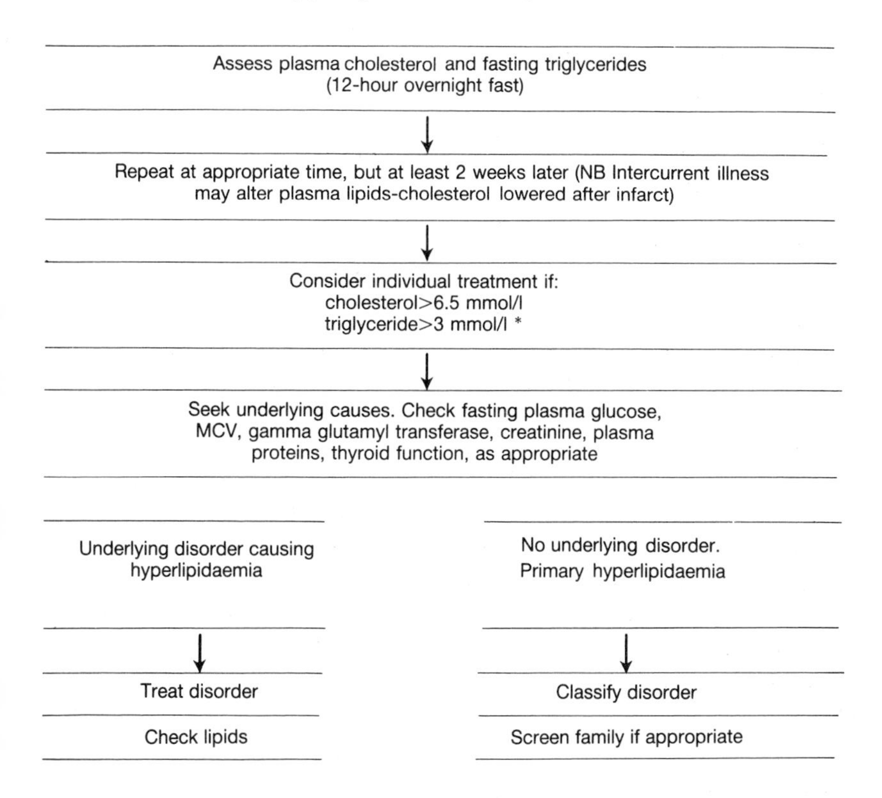

Common causes of secondary hyperlipidaemia:

- Obesity
- Diabetes
- Alcohol excess
- Hypothyroidism
- Renal failure

* European Atherosclerosis Society suggest triglycerides of > 2.3 mmol/l

Fig. 5.2. Plan of investigation of people with hyperlipidaemia.

1. Plasma cholesterol levels tend to increase with age, and are lower in children than in adults.

2. Slightly different values may be obtained when analyses are performed in different laboratories as standardization is difficult. Values

given in subsequent discussion are for general guidance. It is sensible for serial measurements to be performed in one laboratory.

3. A myocardial infarction is often the event which causes the doctor to measure plasma lipids. Plasma lipids are affected by this, and other, severe illnesses. Plasma cholesterol often starts to fall about 24 hours after a myocardial infarct, and may remain at a reduced level for up to two months. Plasma triglycerides may rise, or, occasionally, fall.

4. Plasma cholesterol changes very slightly during the day, but the change in relation to meals is not noticeable. In contrast, plasma triglyceride concentrations are greatly affected by food intake, and specimens should be taken when patients are fasted.

5. Very high lipid levels may cause interference with other laboratory tests (see text and Table 5.1).

When investigating the possibility of hyperlipidaemia in a patient the total cholesterol and fasting triglyceride concentration are the initial investigations. Consideration of the results together with information about the patients such as their history, physical signs, and their family history then usually allows a tentative diagnosis to be made. Measurement of HDL cholesterol and calculation of LDL may be required in children and in those with borderline cholesterol values or high triglyceride levels when it is considered that the result will affect the diagnosis or the decision on treatment. In the current financial climate it would seem a better use of resources to measure total cholesterol on more patients rather than performing lipoprotein fractions as the initial screening procedure.

Lipid concentrations in children

Knowledge about the plasma cholesterol and triglyceride concentrations which should be regarded as healthy for children is still imperfect. The largest epidemiological studies have been made in the USA.

The statistically 'normal' values found in the Bogalusa Heart Study are shown in Table 5.2.

It must, however, be remembered that in line with the situation in adults, those values probably do not represent healthy values, as many children in the Western world consume large quantities of saturated fat.

Up to the age of 5 boys and girls have similar levels, but between 5 and 13 girls tend to have higher LDL and triglyceride levels. After puberty,

Table 5.1. Possible effects of lipaemia on routine biochemical estimations

Analysis	Effect of lipaemia
Electrolytes	Lipids occupy space therefore interfere with the results of electrolyte determinations on flame photometers. Interference is less on an ion specific electrode.
Liver enzymes including aspartate transaminase	Results may be inaccurate, particularly estimations that involve kinetic enzymic assays, because of high 'blank'.
Amylase	May be falsely low by some methods
Calcium	Measurement by fluorescence method may be low due to 'quenching'.
Blood gases	Interference depends on analyser (see manufacturers information). Wash solution should be put through analyser after lipaemic samples to prevent accumulation of lipid on the electrode membrane.
Platelet count	On some counters lipid particles may be counted as platelets

Note: Fat emulsions given in parenteral nutrition may also cause lipaemia. It is wise to take blood 4–6 hours after lipid infusion to allow clearance.

the LDL levels rise in boys and their HDL levels become lower than those of girls, which is the situation in adults. When screening young children for hyperlipidaemia, measurement of the lipoprotein fractions is often needed and the results are best interpreted by someone with experience.

Laboratory measurements

Cholesterol and triglyceride levels are measured enzymatically in most

Table 5.2. Lipid levels in children—mean (5th, 95th centile)

Age	Total cholesterol (mmol/l)	HDL cholesterol (mmol/l)	Triglyceride (mmol/l)
Newborn	1.8	0.9	0.4
(cord blood)	(1.1–2.1)	(0.3–1.5)	(0.1–0.9)
6 months	3.4	1.3	1.0
	(2.3–4.9)	(0.6–2.2)	(0.6–1.9)
1 year	3.9	1.3	0.9
	(2.5–4.9)	(0.6–2.2)	(0.5–1.8)
2–14 years	4.1	1.7	0.7
	(3.1–5.4)	(0.8–2.6)	(0.4–1.4)

laboratories. Cholesterol measures on serum are 3–5 per cent higher than those on plasma.

Observation of serum or plasma stored overnight at 4°C for opalescence (which indicates excess very low or intermediate density lipoprotein) or a creamy upper level of chylomicrons may be helpful.

Measurement of LDL and HDL cholesterol is not required on all patients for routine clinical management, but it is necessary in some cases for diagnosis and for research purposes. The various lipoproteins can be separated by a number of techniques including ultracentrifugation and chemical precipitation. The commonest used methodology is the chemical precipitation of VLDL and LDL, and measurement of the HDL cholesterol left in the supernatant. If triglyceride levels (reflecting VLDL) are normal, then LDL constitutes most of the difference between the total and the HDL cholesterol. Some authors believe that the ratios of the various lipoproteins correlate well with CHD risk, but they may sometimes prove misleading. For example, incorrect risk assessment may be made if the total cholesterol/HDL cholesterol ratio is used without also measuring the triglycerides and estimating the LDL cholesterol.

$$\text{LDL cholesterol} = \text{total cholesterol} - \text{HDL cholesterol} - \frac{\text{triglyceride}}{2.3\ \text{(mmol/l)}}$$

This can only be used if the plasma triglyceride concentration is below 4.5 mmol/l (400 mg/dL). Lipoprotein electrophoresis is not generally helpful, although it is occasionally performed to examine for the presence of a broad B band.

Measurement of lipoprotein lipase is complex, and only performed in a few specialized laboratories, but it may help in the diagnosis of selected cases of severe hypertriglyceridaemia.

Many of the apolipoproteins can now be measured by immunological methods or isoelectric focusing and the usefulness of these measurements is under investigation. Some small retrospective studies performed in specialized centres have shown good correlation between apo B and apo B/ apo A1 and CHD. There is, however, less prospective data or information from large studies such as exists for cholesterol and the lipoprotein fractions. The routine use of these assays is thus not recommended outside specialized centres, at present.

When considering laboratory analysis it is worth noting that severe hyperlipidaemia may cause problems in the determination of some routine blood and plasma analyses. This may sometimes cause difficulty in determining the underlying cause of some cases of severe secondary hyperlipidaemia, and in the biochemical investigation of concurrent diseases. Some of the problems which may be encountered are shown in Table 5.1. Of particular note is the fact that some types of assay for amylase activity are upset by a very high triglyceride level. This may cause a low result and a missed diagnosis of pancreatitis, which may in fact have resulted from the lipid abnormality.

6 Secondary hyperlipidaemia

Diagnosis of secondary hyperlipidaemia

The first step in the management of a patient with hyperlipidaemia is to consider the possibility of an underlying cause, as lipid metabolism is altered in a number of physiological and pathological conditions. Secondary hyperlipidaemia, particularly due to obesity and a diet high in saturated fat, is common. Important causes of secondary hyperlipidaemia are shown in Table 6.1, together with the lipid changes which they tend to cause.

These conditions can usually be diagnosed by a full history and clinical examination, drug history, and determination of fasting plasma glucose, mean corpuscular cell volume, gamma glutamyl transpeptidase, thyroid,

Table 6.1. Conditions which may cause secondary hyperlipidaemia

Condition	Plasma cholesterol	Plasma triglycerides
Diabetes mellitus	(↑)	↑↑
Hypothyroidism	↑↑	↑
Excess alcohol intake		↑↑
Obesity	↑	↑
Nephrotic syndrome	↑↑	(↑)
Pregnancy	↑	
Biliary obstruction	↑	
Myeloma	↑	↑
Acute intermittent porphyria	↑	
Drugs—thiazides		↑
β blockers		↑
steroids	↑	↑
oral contraceptives		↑

renal, and liver function tests, as considered necessary. The response to the drugs mentioned may be general or idiosyncratic (Chapter 10) and a trial of exclusion or alternate medication may sometimes be appropriate. If a hyperlipidaemia is secondary then treatment of the underlying cause is usually associated with the reduction in lipid levels.

In children, noteworthy causes of secondary hyperlipidaemia include diabetes mellitus, hypothydroidism, idiopathic hypercalcaemia, and glycogen storage disease types I and III. Lipid levels rise in the second half of pregnancy.

Diabetes

A significant number of diabetics have elevated plasma triglycerides, particularly when their control is not optimal. Raised plasma cholesterol levels may also occur, especially in type II diabetics. Abnormal lipoprotein composition can be seen even in patients with normal cholesterol and triglyceride levels. Hyperlipidaemia is partly explained by the important role of insulin in both the catabolism of triglyceride-rich lipoproteins, and in LDL receptor activity. Insulin is necessary for the normal activity of lipoprotein lipase, and in the insulin deficiency of severe, uncontrolled diabetes mellitus the effect is of an acquired lipoprotein lipase deficiency resulting in markedly raised plasma triglyceride levels. This can occasionally cause the appearance of eruptive xanthomas (Fig. 6.1), hepatomegaly and lipaemia retinalis. Repletion of insulin restores lipoprotein lipase activity; the triglyceride levels fall and any eruptive xanthomas disappear.

In diabetes there is also a reduction in the peripheral use of glucose. Triglyceride in fat stores is broken down and fatty acids are released into the plasma; some are used by muscle, but a significant rise in blood levels may occur. Increased fatty acid mobilization leads to enhanced VLDL triglyceride secretion from the liver. Metabolism of the fatty acids also results in the formation of ketone bodies.

Glucose intolerance and obesity

The adapted WHO criteria for the diagnosis of glucose intolerance are given in Table 6.2.

In the obese patient with mild glucose intolerance, hyperlipidaemia may be due to triglyceride and VLDL over-production. These patients often have relatively high insulin levels, and several studies have shown a relationship between plasma triglyceride and raised immuno-reactive

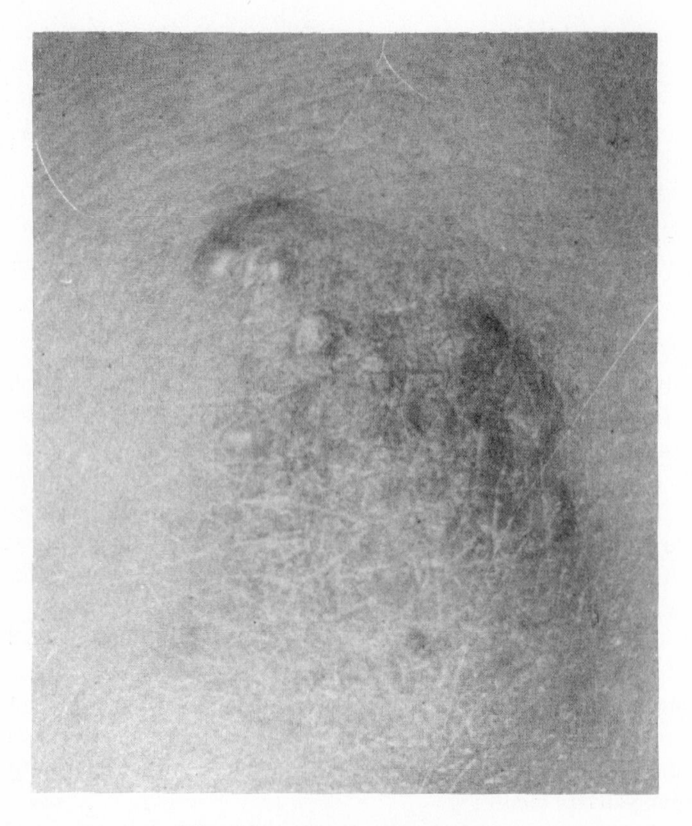

Fig. 6.1. Eruptive xanthomas.

insulin in those patients with impaired peripheral tissue metabolism of glucose. High density lipoprotein levels are generally normal or low.

Weight reduction, which is associated with a fall in plasma insulin, often improves glucose tolerance and reduces triglyceride levels.

Glucose intolerance, gout, and hypertriglyceridaemia are often seen together. Sometimes the cause is diet related, but the problems may persist despite dietary manipulation.

Hypothyroidism

Thyroid hormones appear to enhance cell-surface lipoprotein receptor and lipoprotein lipase activity, and to increase bile steroid excretion.

Table 6.2. Diagnosis of diabetes mellitus and impaired glucose tolerance. Based on the recommendations of the European Diabetes Epidemiology Study Group concerning the second report of the WHO Expert Committee. Diagnostic values for oral glucose tolerance test under standard conditions, using a load of 75 g glucose in 250–350 ml of water for adults, and of 1.75 g/kg body weight (to a maximum of 75 g) for children, and specific enzymatic glucose assay. Two classes of response are identified—diabetes mellitus and impaired glucose tolerance

Glucose concentration (mmol/l)	Venous whole blood	Capillary whole blood	Venous plasma
Diabetes mellitus			
Fasting	≥6.7	≥6.7	≥7.8
and/or			
2 hours after glucose load	≥10.0	≥11.0	≥11.1
Impaired glucose tolerance			
Fasting	<6.7	<6.7	<7.8
and			
2 hours after glucose load	6.7–9.9	7.8–11.0	7.8–11.0

The WHO Expert Committee recommended the following procedure for diagnosis:

1. If symptoms of diabetes are present, perform random or fasting blood glucose measurement. In adults, random venous or plasma values of 11 mmol/l or more or fasting values of 8 mmol/l or more are diagnostic. Random values below 8 mmol/l and fasting values below 6 mmol/l exclude the diagnosis.
2. If results are equivocal, measure blood glucose concentration two hours after 75 g of glucose taken orally after an overnight fast. Two-hour venous plasma glucose values of 11 mmol/l or more are diagnostic of diabetes. Values below 8 mmol/l are normal and those in the range 8–11 mmol/l are termed impaired glucose tolerance.

Hypothyroidism results in a rise in circulating LDL and total cholesterol levels. It may also cause hypertriglyceridaemia because of the reduction in hepatic lipase activity, despite the fact that the supply of fatty acids to the liver is less due to reduced triglyceride breakdown in adipose tissue. Hypothyroidism may also result in a marked hyperlipidaemia if it occurs in genetically-disposed people (see remnant hyperlipidaemia, Chapter 7). There is usually a rapid fall in lipid levels in response to treatment with thyroxine.

Chronic renal failure

Raised levels of plasma triglycerides, or of triglycerides, LDL, and IDL cholesterol, may occur in chronic renal failure. This may be due to the reduced activity of hepatic lipase and lipoprotein lipase. VLDL production may also be increased, and HDL levels tend to be low. Cardiovascular disease is one of the most frequent complications in patients receiving haemodialysis, and hyperlipidaemia may be an important risk factor in these patients. The lipoprotein abnormalities often disappear after successful renal transplantation, although changes may persist if the patient has to remain on high doses of corticosteroids because they reduce lipoprotein lipase activity.

Liver disease

Severe liver disease is accompanied by various disorders of lipid metabolism. Esterification of cholesterol is low because of reduced activity of the responsible enzyme. The lipoproteins often have an abnormal composition—and the abnormal lipoprotein X may be found. Raised plasma triglycerides are probably caused by low hepatic triglyceride lipase activity, but a triglyceride-rich LDL fraction may also contribute. HDL levels are often low.

All these changes are usually fully reversible and seem to be parameters indicating the degree of disturbance of liver function during inflammatory liver disease. Underlying alcohol abuse must be considered as this is notoriously under-reported.

Alcohol excess Alcohol causes a rise in plasma triglycerides primarily by inhibiting fatty acid activation and increasing fatty acid synthesis in the liver. *VLDL* output from the liver goes up. The amount of alcohol required to cause a significant rise in plasma levels depends on the individual—15–20 units per week may be enough in a susceptible person. The lipid levels often fluctuate. Some people can develop severe hypertriglyceridaemia and even chylomicronaemia after excessive alcohol intake. Pancreatitis can occur as a serious complication in such individuals where plasma triglyceride levels exceed 20 mmol/l.

Nephrotic syndrome

The causes of the hyperlipidaemia which often occur in this condition are unclear. In nephrotic syndrome there is an increase in hepatic protein synthesis as a response to the high urinary losses. Over-production of the lipoproteins VLDL, and probably LDL, also seems to occur and the plasma levels are usually raised. HDL cholesterol and apoA1 levels may be very low if urinary losses are high. It is of note that it is the HDL subfraction, HDL_2 (which is inversely related to CHD risk), which is particularly reduced. Another suggested mechanism for the lipoprotein changes is that hypoalbuminaemia causes fatty acids to bind to low affinity sites and increases fatty acid uptake by tissues including the liver, and this may increase VLDL synthesis. A further possibility is that hypoalbuminaemia causes an increase in apoprotein B synthesis.

It is not clear whether the CHD risk associated with secondary hyperlipidaemia is as high as might be expected considering the lipid levels. One factor which could be expected to explain a different risk from the inherited primary conditions is the duration of the hyperlipidaemia, which is generally shorter. Pancreatitis is, however, a significant risk in secondary hypertriglyceridaemia as in the primary conditions.

A plan for the identification of secondary hyperlipidaemia is Fig. 6.2. For many of the conditions causing secondary hyperlipidaemia, such as hypothyroidism, effective treatment of the condition results in a reduction in plasma lipid levels. The response should, however, be checked because primary hyperlipidaemia is common and people may have this as well. In secondary cases, where treatment of the underlying condition is not possible or does not result in a reduction of lipid, as in some renal disorders, then specific treatment to lower the lipid levels may be necessary. In these cases the management required may be different from that which would usually be prescribed. For example, the diet advised for a patient with nephrotic syndrome and severe hyperlipidaemia will need to be specially designed for the individual patient. Consideration of drug treatment will also need to take into account the

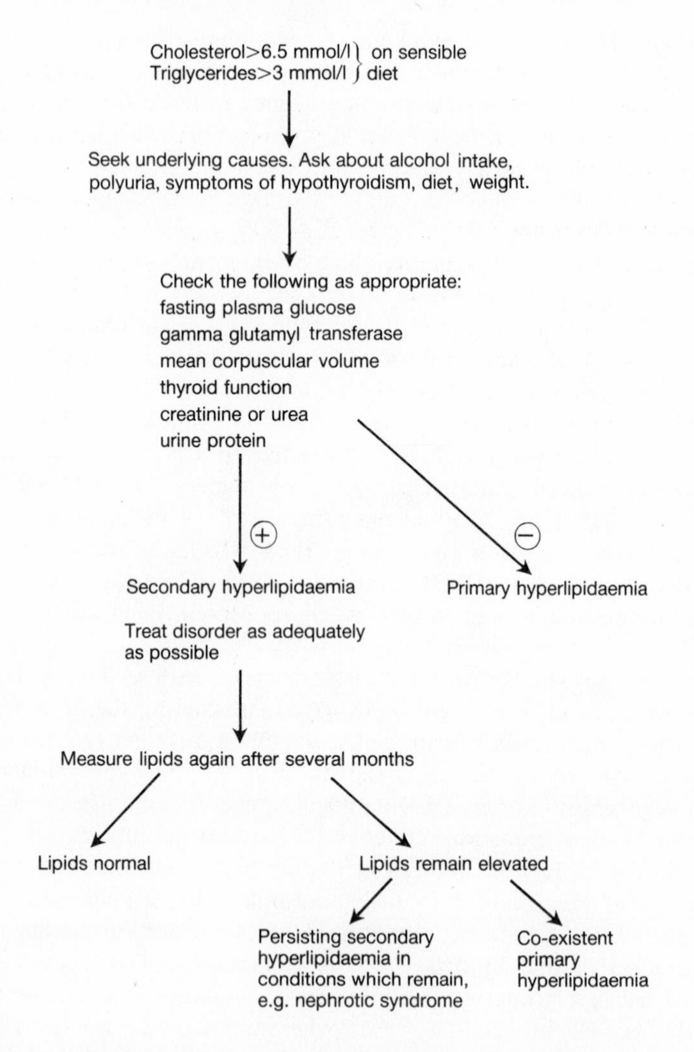

Fig. 6.2. Schematic diagram of investigations which should be considered in patients with hyperlipidaemia.

metabolism of the drug in the particular disease, for example bezafibrate must be given in reduced doses if it is used in patients with renal impairment.

Summary

Identification of secondary hyperlipidaemia is the first step in the assessment of any lipid abnormality. Conditions such as hypothyroidism and alcohol excess, which require specific treatment in their own right, may be identified because of the finding of hyperlipidaemia. Measurement of plasma lipids is also important in diabetics as they are at particular risk of developing CHD and vascular problems.

7 Primary hyperlipidaemia

Once secondary hyperlipidaemia has been excluded, then characterization of the disorder is needed. A precise diagnosis is essential since these disorders have major implications concerning prognosis, genetic counselling, and life-long diet and drug treatment.

Several classifications have been suggested for the primary hyperlipidaemias. The simplified genetic metabolic classification given in Table 7.1, based on that introduced by Goldstein *et al.*, is clinically more relevant than the World Health Organization (Fredrickson) classification. The Fredrickson classification is based on the patterns seen on lipoprotein electrophoresis and these are not specific to disease entities.

The distinct genetic conditions which are currently recognized are described in this chapter. It is likely that as our knowledge increases further groups will be delineated. At present those who have primary hyperlipidaemia, but do not obviously fit a condition described, are usually said to have 'common' hyperlipidaemia.

Familial hypercholesterolaemia

In this disorder there is a high concentration of LDL and total cholesterol in the circulation because of an inherited defect of LDL metabolism. The defect is usually a marked reduction in the number of the high-affinity LDL receptors on the surface of cells, which results in less LDL being removed from the circulation, but other defects such as functionally defective receptors and reduced receptor binding may give a similar picture.

Familial hypercholesterolaemia (FH) is inherited in an autosomal dominant manner and is one of the commonest inherited conditions. In Britain, it is estimated that at least 1 in 500 people are affected, but in some populations, such as the Lebanese and South Africans of Afrikaans descent, the incidence is much higher.

FH is a very serious disorder, with a much greater risk of development of premature CHD than common hyperlipidaemia: a fact which is partly due to elevated LDL from birth. About 1 in 20 patients presenting with a myocardial infarction under age 60 have this condition. Untreated, up to

Table 7.1. The Goldstein and Fredrickson classification of primary hyperlipidaemias

Characteristics of primary hyperlipidaemia [1] ↑ =raised; ↑ ↑ =markedly raised; N=Normal

	Atherosclerosis risk	Xanthomas	Inheritance	Relative prevalence	Lipid abormalities	WHO type
Familial hypercholesterolaemia						
	+++	Tendon	Autosomal dominant	++	Cholesterol and LDL↑↑ Triglyceride N or slightly ↑	IIa, occasionally IIb
Familial combined hyperlipidaemia						
	++	–	?	+++	Cholesterol and LDL ↑ Triglyceride and VLDL↑	IIb, occasionally IIa and IV
Remnant hyperlipoproteinaemia (broad beta)						
	++	Tuboeruptive palmar	Apo EIII deficiency and other factors	+	Cholesterol ↑ ↑ Trigylceride ↑ ↑ IDL ↑	III
Familial hypertriglyceridaemia						
Excessive synthesis	?	Eruptive	Probably autosomal dominant	+	Triglyceride and VLDL ↑↑ Chylomicrons ↑	IV, V
Lipoprotein lipase/apo CII deficiency	?	Eruptive	Probably recessive	rare	Triglyceride ↑ ↑ Chylomicrons ↑ ↑ VLDL ↑	I, V
Common hypercholesterolaemia						
	+		Polygenic	++++	Cholesterol and LDL ↑	IIa

[1]Adapted from Lewis B. Disorders of lipid transport. In: *The Oxford textbook of medicine,* Weatherall DJ. Ledingham JGG, Warrell DA, eds. Oxford: Oxford University Press, 1983: 9.58–9.70.

85 per cent of men with FH will have a myocardial infarction and 50 per cent will have died by the age of 60 years. Patients may be disabled by CHD in their 30s. The family history below illustrates the devastating effects it can have (Fig. 7.1).

Women with FH seem to have a better prognosis than men, with CHD developing some 10–15 years later.

FH is characterized by xanthomas in the extensor tendons of the hands, the Achilles tendons (Figs 7.2 and 7.3), and, occasionally, at the

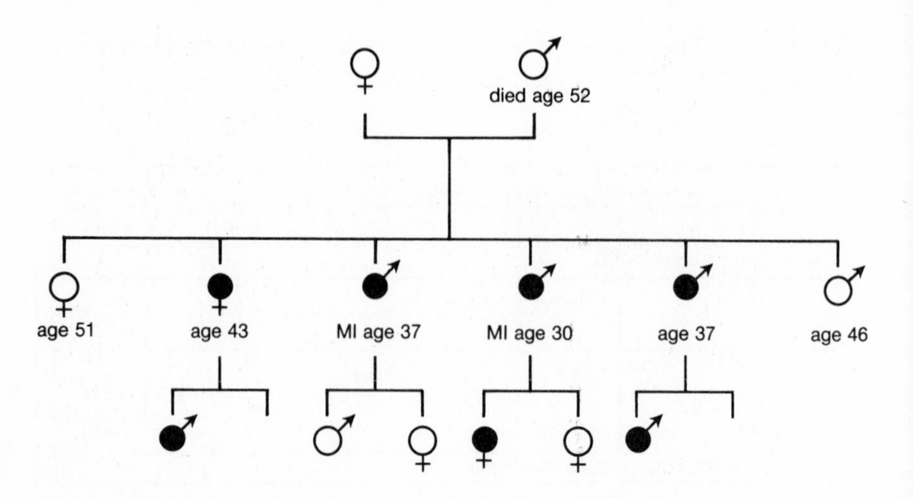

Affected adults all had tendon xanthomas and a plasma cholesterol>9 mmol/l at diagnosis. The family was screened after patient X had a myocardial infection at age 30 and was found to have a cholesterol of 13 mmol/l. Screening has allowed diagnosis in six other family members and commencement of cholesterol-lowering therapy.

Fig. 7.1. Family tree showing individuals with familial hypercholesterolaemia.

insertion of the patella tendon. Corneal arcus commonly occurs in FH, though it may occur in other types of hypercholesterolaemia and is frequently seen in old people with normal lipids. Half arcus in a young person is, however, very suggestive of FH (Fig. 7.4). Xanthelasma (Fig. 7.5) are less specific clinical signs, because they may be found in people with a normal cholesterol and in those with common hyperlipidaemia.

Total cholesterol levels in heterozygous patients may range from 7 to 15 mmol/l, but are commonly between 8.5 and 12 mmol/l in affected adults. Levels in affected children are usually somewhat lower.

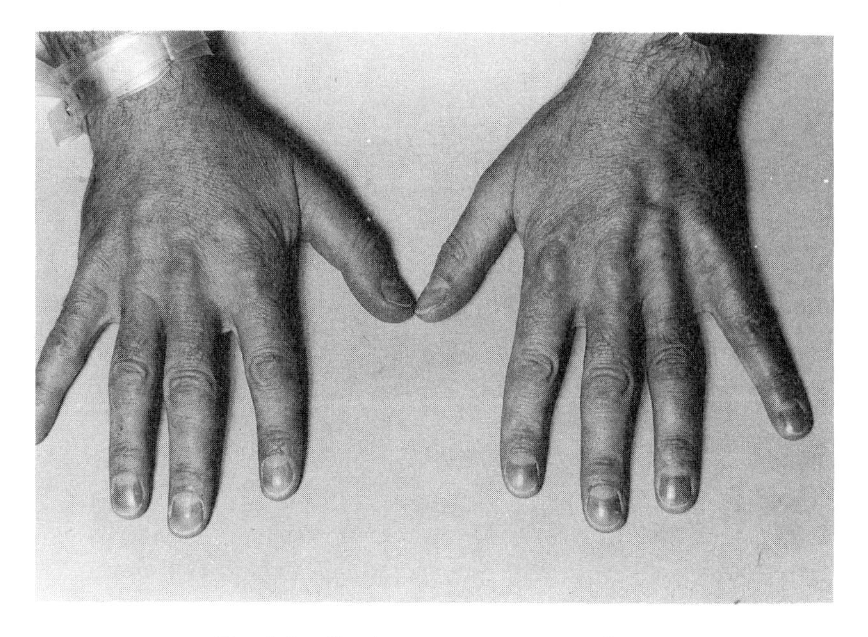

Fig. 7.2. Xanthomas in extensor tendons.

Triglyceride and VLDL levels are normal or slightly raised and HDL cholesterol is usually decreased. The presence of xanthomas and elevated levels of total plasma cholesterol (chiefly found in LDL) are sufficient evidence to make the diagnosis. The xanthomas may, however, be slight and must be carefully sought. If xanthomas are absent, as they often are in childhood or early adult life, the condition may be diagnosed if LDL cholesterol is greater than 5 mmol/l and FH has been diagnosed in a first-degree relative. It is usually possible to diagnose this condition in children, although LDL should be measured in most cases because both the total and the LDL cholesterol are lower in children (see Chapter 5). Neonatal diagnosis can be made from cord blood, but it is more common for the results to be inconclusive when measures are made at this time and a repeat test is usually performed later. Genetic counselling is discussed in Chapter 10.

Energetic treatment is essential for people with FH. Dietary modification is considered in Chapter 8, and this may result in a modest reduction of cholesterol levels, particularly if the patient was obese. However,

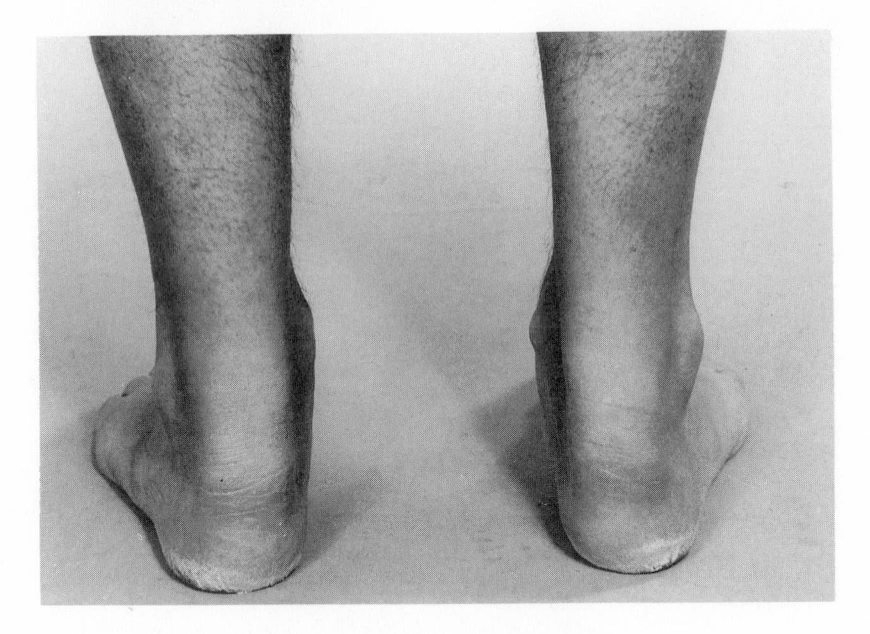

Fig. 7.3. Achilles tendon xanthomas.

diet alone seldom achieves satisfactory lipid levels and lipid-lowering drugs are necessary. Bile acid sequestrants, cholestyramine or colestipol, are often used, but fibric acid derivatives are possible additional or alternative drugs, and HMGCoA reductase inhibitors are now being used. In some centres ileal bypass surgery has been performed in resistant cases, but this procedure has frequently resulted in severe gastrointestinal side-effects and is now very seldom performed. All these cholesterol-lowering treatments are associated with a reduction in size of xanthomas. Clinical trials including patients with FH indicate benefit in terms of morbidity and mortality.

In the rare homozygous form of FH the patients have virtually no LDL receptors on their cells and the plasma cholesterol levels are often greater than 20 mmol/l. Thus, xanthomas, aortic stenosis, and CHD may occur in childhood. Drug treatment seldom achieves normal lipid levels, and plasma exchange together with combined drug treatment appears to be the therapy of choice. Even with treatment, the prognosis is poor and few survive beyond the age of 20.

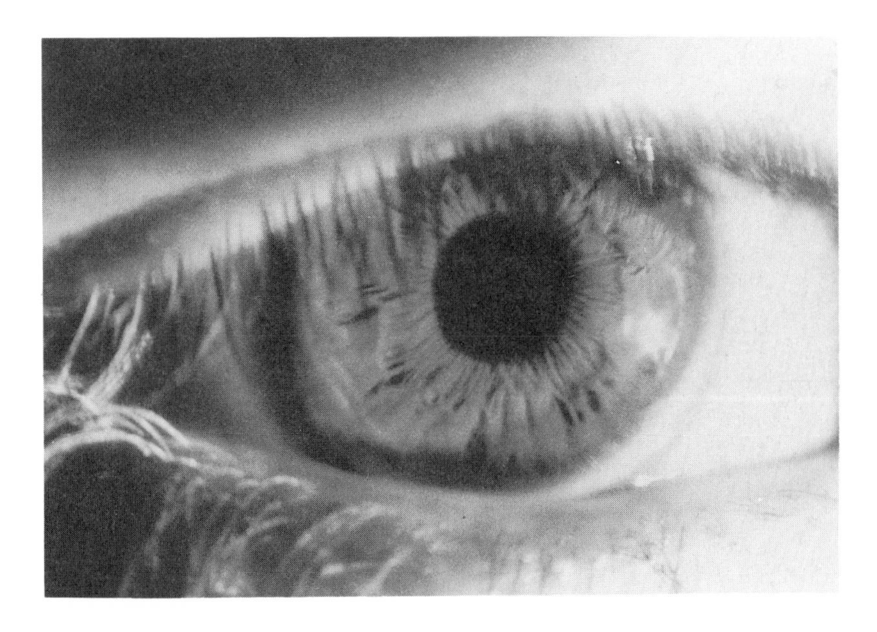

Fig. 7.4. Corneal arcus in a 35-year-old.

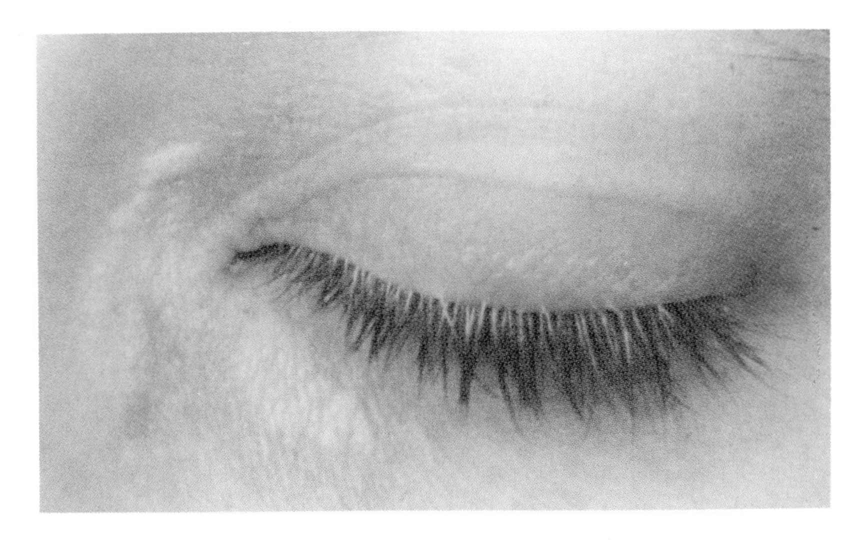

Fig. 7.5. Xanthelasma.

Table 7.2. Criteria for diagnosis of familial hypercholesterolaemia (FH) used for the Familial Hyperlipidaemia Register—a British register of patients with this condition

Definite FH			
Cholesterol	>7.5 mmol/l (Adult)		
or	>6.5 mmol/l (Child under 16)		
LDL cholesterol	>4.9 mmol/l (Adult)		
Plus			
Tendon xanthomas in patient or relative			
Possible FH			
Cholesterol	>7.5 mmol/l (Adult)		Family history myocardial infarction
	>6.5 mmol/l (Child under 16)		under 50 (2° relative)
		Plus	under 60 (1° relative)
or			or
			Family history raised
LDL cholesterol	>4.9 mmol/l (Adult)		cholesterol in 1° relative

Possible familial hypercholesterolaemia

Primary hypercholesterolaemia may be found in several family members although none have tendon xanthomas. Without performing receptor studies, which is impractical, it is not possible to tell whether each family has a receptor defect or whether there is a different inherited metabolic defect. In some families it is difficult to elicit a family history suggestive of an inherited hyperlipidaemia. This is particularly the case if a disorder is passed from the women in a family, as death from another disease such as cancer, may occur before the women develop clinical CHD.

Table 7.2 shows the FH register criteria for the diagnosis of FH and possible FH. The category of 'possible FH' is unfortunately likely to include a number of cases of familial combined hyperlipidaemia.

Familial combined hyperlipidaemia

This inherited disorder was only recognized as a separate condition in the 1970s and it is still incompletely understood. It does, however, appear to be common with an incidence which is as high as 1 in 300. The precise mode of inheritance has not been resolved, but it is probably autosomal dominant. It is characterized by elevated levels of VLDL and often of LDL (and as a result, both cholesterol and triglyceride) and low levels of HDL. The high VLDL is generally due to increased synthesis; and apoB levels are generally high due to increased hepatic synthesis. Cholesterol levels are usually 8–10 mmol/l and triglycerides 3–6 mmol/l. This condition should be suspected in families with a strong history of premature CHD and the characteristic metabolic disturbance in several family members, but no xanthomas. The expression of the metabolic abnormality may differ in members of the same family, probably as a result of interaction between genetic and environmental factors, especially diet. Some family members may have predominant hypertriglyceridaemia, some hypercholesterolaemia, and the pattern may also vary with age. Unlike FH, the disorder is not present until adult life and in small families the distinction between this condition and 'common' hyperlipidaemia may be difficult and sometimes impossible. It is nevertheless important to attempt to establish the diagnosis since the prognosis of familial combined hyperlipidaemia is much worse than that of common hyperlipidaemia. The hyperlipidaemia may sometimes respond to dietary modification alone, but if it does not the fibrate drugs may be used, sometimes in combination with a bile acid sequestrant.

Remnant hyperlipoproteinaemia

In this condition elevated levels of cholesterol (around 10–15 mmol/l) and triglyceride (5–12 mmol/l) are due to accumulation of IDL and the denser sub-classes of VLDL, as well as some chylomicrons. LDL and HDL levels are low. VLDL in this condition has a number of atypical features: slow electrophoretic mobility (giving a broad B band), a high ratio of cholesterol to triglyceride (a molar ratio greater than 1:1 in the VLDL fraction) and a high content of apo E. Although lipoprotein electrophoresis is now rarely performed, this condition is still sometimes called broad beta disease, or type III hyperlipidaemia. The disorder appears to be due to impaired lipoprotein catabolism which results in an accumulation of remnant particles.

The genetics of this condition are complex. Most people have an apo E_2E_2 genotype and do not produce apo E_3, but as this phenotype is present in 1 per cent of a healthy population, other factors must be involved. It has been suggested that apo-E_3 deficiency may interact with other inherited hyperlipidaemias (e.g. FH, familial combined), hypothyroidism, or diabetes to produce remnant hyperlipidaemia. While the genetics and mechanism may be complicated, the clinical manifestations are clearly defined. Many patients have the typical cutaneous lesions; linear, orange coloured, planar xanthomas seen in the palmer creases and/or tubo-eruptive xanthomas, which are commonly found on the elbows and which are diagnostic (Fig. 7.6).

Although the precise risks have not been quantified there is no doubt about the increased vascular risk. Peripheral vascular, cerebrovascular disease, and CHD commonly occur. The diagnosis is usually suspected because of the skin lesions and confirmed by finding combined hyperlipidaemia (a similar elevation of cholesterol and triglyceride) and a broad beta band on lipoprotein electrophoresis. Occasionally further tests such as apo-E typing and confirmation of a VLDL cholesterol:triglyceride molar ratio >1 may be needed. The xanthomas and metabolic abnormality may respond promptly to the treatment of any secondary factors, but if not then they usually respond rapidly to diet and clofibrate or fibrate analogues.

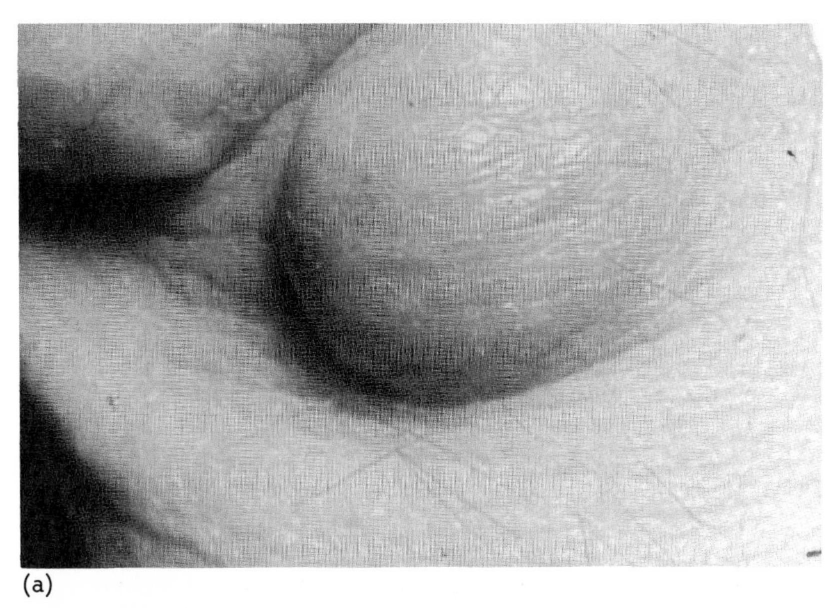

(a)

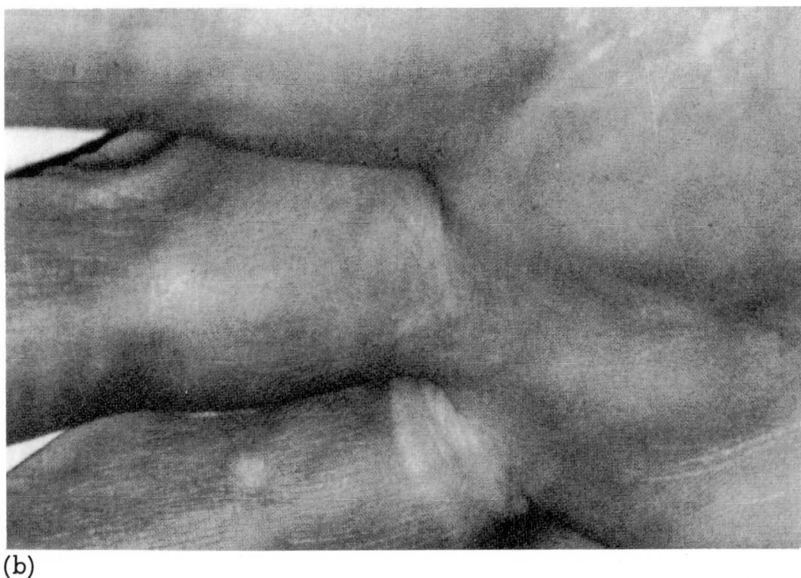

(b)

Fig.7.6. (a) Tubo-eruptive xanthomas and (b) palmar crease xanthomas. (Courtesy of Professor B. Lewis.)

Familial hypertriglyceridaemia

There are several inherited forms of hypertriglyceridaemia. If the triglyceride levels are very high, say over 20 mmol/l, then there may be the classical clinical features of eruptive xanthomas (Fig. 7.7), lipaemia retinalis, abdominal pain, hepatosplenomegaly, and an increased risk of acute pancreatitis. The association with CHD is less clear and may depend on other factors such as HDL cholesterol; being more significant if the HDL cholesterol is low.

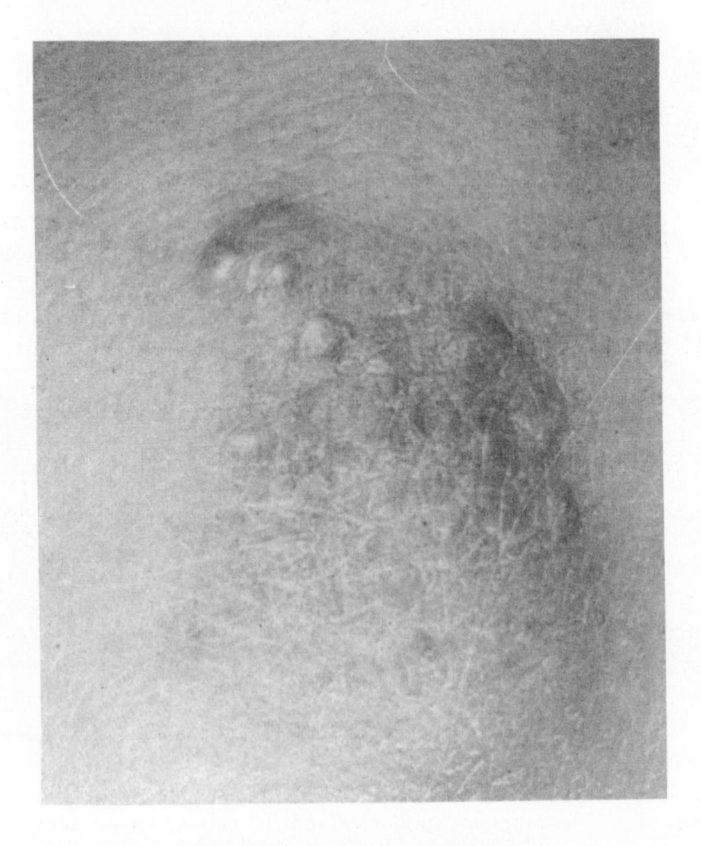

Fig. 7.7. Eruptive xanthomas.

Endogenous hypertriglyceridaemia

The main abnormality is an increase in VLDL probably due to excessive synthesis. There may also be chylomicronaemia which disappears with a fat-free diet while the VLDL abnormality persists. Triglycerides range from 5 to 50 mmol/l. Cholesterol is only modestly increased and LDL and HDL levels are low. Affected relatives show a consistent metabolic abnormality, and impaired glucose tolerance, or diabetes and obesity, are often associated. The inter-relationships are not clearly understood, but hyperinsulinaemia may be relevant. Plasma stored overnight in the refrigerator shows diffuse opalescence, with or without a creamy upper layer of chylomicrons. Dietary modification is relatively ineffective. Clofibrate, bezafibrate, nicotinic acid, and progestogens have been used with varying success.

Lipoprotein-lipase or apo-CII deficiency

A deficiency of lipoprotein lipase or its activator apo CII are very uncommon conditions which lead to severe hypertriglyceridaemia characterized by an inability to clear ingested fat. The inheritance is autosomal recessive and a common history in affected individuals is of recurrent episodes of abdominal pain or pancreatitis and/or eruptive xanthomas. Stored plasma shows a creamy upper layer of chylomicrons. LDL and HDL levels are low. Plasma triglycerides may be elevated from birth and may be as high as 20–100 mmol/l. Lipoprotein lipase deficiency is identified by specific assay available in one or two specialized laboratories. Apo CII can be measured, but a deficiency can be indicated simply by demonstrating a substantial reduction in the hypertriglyceridaemia after the infusion of fresh frozen plasma, which provides the apoproteins. Virtual removal of fat from the diet for three days in hospital, results in a marked fall in triglyceride levels and the chylomicrons often disappear. Long-term treatment involves severe restriction of all fat-rich foods. Medium chain triglycerides may be used.

These conditions are very rare: accumulation of chylomicrons (the 'chylomicronaemia syndrome') is more commonly seen as a secondary phenomenon.

Common 'polygenic' hyperlipidaemia

A considerable number of people with hyperlipidaemia of moderate severity, despite sensible diet, do not have a family history of hypercholesterolaemia or premature CHD. The cause in these individuals may be multi-factorial and involve genetic and environmental factors which have not yet been elucidated. At present it is a diagnosis made after the exclusion of other causes. People with common hypercholesterolaemia do not have xanthomas. The first line of treatment is a lipid-lowering diet. The need for any drug treatment will depend on the patient's age, condition, and lipid levels. More active treatment is usually considered if the individual has other risk factors for CHD.

Familial lipoprotein deficiency

These disorders result not in high but in low lipoprotein levels.

The inherited lipoprotein deficiency disorders are of two major types—those affecting lipoproteins containing apolipoprotein B (chylomicrons, VLDL, and LDL), and those affecting lipoproteins containing the A apolipoproteins (HDL). The disorders may result in either low or unmeasurable concentrations of the apoproteins and the particular lipoprotein particles. Low levels of apo A and HDL are increasingly being recognized in some families with premature CHD. More work is needed on these conditions, but such a problem should be considered in families with a strong history of CHD and none of the more commonly recognized risk factors.

Other lipoprotein deficiency states—such as Tangier's disease and hypo- and abeta-hypolipoproteinaemia—are rare. The clinical features of the three major types are summarized in Table 7.3.

The disorders in which the apo B concentration is low are usually associated with a low incidence of CHD. However some disorders where there is also a low apo A and a very reduced HDL cholesterol may be associated with an increased risk of CHD, even if the LDL cholesterol is relatively low. This is the case in Tangier's disease. This condition is rare, but there is a notable finding of markedly enlarged, orange tonsils which may cause respiratory obstruction.

Table 7.3. Hypolipoproteinaemias

Type	Onset (Inheritance)	Lipids	Clinical effects
Abeta	Early childhood (autosomal recessive)	Cholesterol 1–2 mmol/l Tg <0.2 mmol/l No apo B	Fat malabsorption, ataxia, neuropathy, retinitis pigmentosa
Hypo-beta	Childhood/adult (autosomal dominant)	Cholesterol 1–3 mmol/l Low apo B	Some degree of fat malabsorption
Tangier's	Childhood (autosomal recessive)	Cholesterol 1–3 mmol/l Tg normal HDL cholesterol very low	Large orange tonsils, corneal opacities, poly-neuropathy. ?increased CHD risk
Hypo-alpha	Child to adult	HDL cholesterol low Apo A low	Increased CHD risk

Summary

Primary hyperlipidaemia is a common condition which reduces life expectancy and has devastating effects in some families. There are a number of different disorders and it is important to make an accurate diagnosis because there are serious implications for both the individual and for his/her family.

8 Dietary management of hyperlipidaemia

In recent years there have been several important reports concerning diet and cardiovascular disease. These are mainly concerned with general advice for the population—the healthy eating plan—and are discussed in Chapter 11. Stricter dietary control than that recommended for the general population is often required for people with diagnosed hyperlipidaemia. This can be introduced in two stages. The reduction in energy from fat to 30 per cent or less of the total can be considered as stage A. Some people will be deemed to need the addition of drug therapy while continuing on the stage A diet. Some people may need, and accept, a further reduction which can be considered as stage B.

The main dietary principles

There are several complementary dietary changes which help to reduce cholesterol. Attention to all of them will achieve the greatest cholesterol reduction, but the emphasis needs to be varied, taking into account the patient's preferences and needs.

The main aspects are:

1. Attainment of ideal body weight.
2. Reduction of saturated fatty acids, partially compensated for by an increase in poly- and mono-unsaturated fatty acids.
3. Increase in fibre-rich carbohydrate.
4. Substantial reduction in alcohol intake in overweight patients or those with hypertriglyceridaemia.

Modification of dietary protein source and cholesterol intake may also be helpful.

Attainment of ideal body weight

Obesity is a major cause of hypertriglyceridaemia and weight loss can often produce an appreciable reduction in triglyceride levels. Cholesterol levels may also be reduced to some extent by weight reduction. The aim should be to achieve a body mass index (weight/height^2) of 20–25. Table 8.1 shows the approximate weight range which corresponds to this in men and women of different height. For people who are substantially overweight it is important to set intermediate goals since people often become disheartened if they feel that they have an impossible target. A weight loss of 2 kg per month should be regarded as perfectly satisfactory and 4 kg per month as excellent. Many patients do not need to count calories—changing from inappropriate to appropriate foods often achieves a satisfactory weight reduction. When this is not achieved it is essential for the dietitian (or other appropriately qualified person) to assess approximate energy intake and to advise as to how this might be reduced by about one third. Recommendations of fixed energy diets (e.g. 800 or 1000 kcal reducing diets) can be inappropriate and disheartening, especially for those whose present energy intake is very high.

Reduction of saturated fatty acids and increase in monounsaturated and polyunsaturated fatty acids

Many people have as much as 40 per cent of their total energy as fat. This should be reduced so that fat provides no more than 30 per cent total energy and saturated fatty acids (SFA) no more than 10 per cent. The actual weight of fat which corresponds to this (and to the higher value of 35 per cent and the present average of 42 per cent) for different total calorie intakes is shown in Table 8.2. The reduction in SFA can be partially compensated for by an increase in monounsaturated fatty acids (MUFA) and polyunsaturated fatty acids (PUFA). A rich source of MUFAs is olive oil. Practical examples of possible exchanges are shown in Table 8.3. If accompanied by weight reduction, where necessary, this dietary change will produce an appreciable reduction in plasma cholesterol, LDL, and triglycerides. Fish appears to be valuable in reducing CHD risk when it is part of a low saturated fat diet. The very long-chain polyunsaturated fatty acids from fatty fish help to lower VLDL, although they will only tend to lower LDL when given in very large quantities. An increase in fish consumption is usually advised, particularly to replace meat, but the widespread use of expensive refined supplements is not (see Chapter 9).

In people with an inability to clear ingested fat (those with lipoprotein lipase or apo CII deficiency) it is necessary to reduce fat intake even further—to 10–20 g/day. Such diets usually need supplementation with

Table 8.1. Guidelines for body weight (based on Bray, 1979, Metropolitan Life Insurance Tables, 1960 and Report on Obesity, Royal College of Physicians, London 1984). The minimum level for diagnosing obesity is taken as 20 per cent above the upper limit of the acceptable weight range

Height without shoes		Weight without clothes		
Ft. ins.	Metres	Acceptable average	Acceptable range	Obese
MEN				
5′3″	1.60	58	52–65	78
5′4″	1.63	59	54–67	80
5′5″	1.65	60	55–69	82
5′6″	1.68	62	56–71	85
5′7″	1.70	64	58–73	88
5′8″	1.73	66	59–74	89
5′9″	1.75	67	61–76	91
5′10″	1.78	69	65–80	96
6′0″	1.83	73	66–83	100
6′1″	1.85	75	69–86	103
6′2″	1.88	77	71–88	106
Body mass index		22	20.1–25	30
WOMEN				
5′0″	1.52	48	44–57	68
5′1″	1.55	50	45–58	70
5′2″	1.57	51	46–59	71
5′3″	1.60	53	48–61	73
5′4″	1.63	55	49–63	76
5′5″	1.65	56	51–65	78
5′6″	1.68	58	52–66	79
5′7″	1.70	60	53–67	80
5′8″	1.73	62	55–69	83
5′9″	1.75	63	57–71	85
5′10″	1.78	65	58–73	88
Body mass index		20.8	18.7–23.8	28.6

Conversion: 8 stone = 51 kg, 9 stone = 57 kg, 10 stone = 64 kg, 11 stone = 70 kg

Table 8.2. Weight composition of diets of different fat and calorie content

	Percentage energy provision as fat:					
	30%		35%		42%	
K Cals	Grams of fat	Other* in grams	Fat (g)	Other (g)	Fat (g)	Other (g)
1000	35(10)	167	38	155	46	138
1500	49(15)	250	57	232	68	207
2000	65(20)	333	76	310	92	276
2500	85(25)	417	95	387	114	345
3000	98(30)	500	114	464	137	414
% of total food weight	16%	84%	19%	81%	25%	75%

*Other = carbohydrate, protein and alcohol

30g = 1 oz

[ ] = weight of saturated fat in accordance with recommendation of only 10 per cent of energy total energy intake.

medium-chain triglycerides (MCT) and vitamins and are usually only recommended by dietitians in specialist clinics.

Increase in fibre-rich carbohydrate

Carbohydrates are increased to compensate for the reduction in fat and it is appropriate that most of the carbohydrate should be unprocessed and rich in dietary fibre. While insoluble fibre (i.e. bran and other fibre from cereal sources) may be of general benefit and helpful in reducing some forms of gastrointestinal disease, it is only the insoluble or gel-forming fibres (e.g. those in various cooked dried beans, oats) and pectins (in fruit) which are of particular value in lowering LDL levels. Refined carbohydrate intake needs to be reduced as it may have an adverse effect if an individual has mild glucose intolerance, and even people with a normal glucose tolerance test may show a fall in triglyceride levels when sugar intake is specifically reduced. All sugars should thus be severely restricted in those with hypertriglyceridaemia and in those who are overweight. Contrary to the belief of some, brown sugar is no better than white, and it is always better that sugars in the diet be taken in the 'natural' form (fruits and milk) rather than in processed foods.

Alcohol

There is epidemiological evidence that for most of the population a modest alcohol intake may be beneficial or neutral. Some people are, however, particularly sensitive to its effects on plasma lipids, body weight, and blood-pressure. Moderate quantitites of alcohol tend to increase triglyceride levels. Alcohol must therefore be severely restricted or preferably eliminated from the diet of people with hypertriglyceridaemia and those who are appreciably overweight. People with other disorders should probably restrict their intake to about 12 units per week.

Protein

Vegetable proteins derived from legumes (e.g. soya) may help to lower LDL cholesterol.

Salt

In view of the fact that dietary sodium can influence blood-pressure it is probably appropriate to advise modest sodium restriction to patients following a lipid-lowering diet. This is in keeping with the general recommendations for healthier eating, and in general just involves avoiding adding salt at the table.

Dietary cholesterol

Dietary cholesterol (from eggs, shellfish, and dairy products) increases blood cholesterol when taken in large quantities by people who have a relatively high intake of saturated fat. Reducing fat intake will in itself reduce dietary cholesterol and also seems to reduce the effect of dietary cholesterol on blood cholesterol. On a diet reduced in fat, intake is usually below 300 mg per day and there is no evidence that restricting dietary cholesterol further gives additional benefit.

Table 8.4 summarizes the optimal nutrient composition of a lipid-lowering diet.

Table 8.3. Total and polyunsaturated fat (PUFA) content of various foods by wet weight and straightforward low saturated fat alternatives.

Product	FAT (grams/100 g wet weight)		Alternative	FAT (g/100 g)	
	Total	PUFA		Total	PUFA
Whole milk	3.8	0.1	Skimmed milk	0.1	—
			Semi-skimmed	1.5	0.1
Salad cream	27		Low calorie salad cream	13	
Double cream	48	1	Yoghurt	1	—
Cream cheese	47	1	Cottage cheese	4	
Cheddar cheese	38	1	Edam/camembert	23	1
Stilton	40		'Low fat cheeses'	15–25	1

Butter	82	2	'High PUFA' margarines	81	50–60
Hard margarine	81	14	Low fat spread	41	12
Lard	99	9	Sunflower oil	100	52
Coconut oil	90	2	Soya oil	100	57
Grilled streaky bacon	36	3	Roast chicken	5.5	1
Grilled beefburger	20	1	Rabbit – stewed	8	3
Sausages (pork)	25	2	Lean ham	5	1
Pork chop – grilled	24	2	Kidney	5	1
Luncheon meat	27	2	Tuna fish (in veg oil)	22	8
Pork pie	27	2	Grilled white fish	1.5	0.5

Note: All figures are an approximate value (based on tables of McCance and Widdowson in *The Composition of Foods* (4th edn) HMSO).

Note: These figures should be used to compare the percentage fat of a weight of similar products when considering substitution. They do *not* provide a guide to the percentage of energy from that food which will be from fat or saturated fat.

Table 8.4. Summary of the optimal nutrient composition of a lipid-lowering diet

Total energy Should be tailored to individual requirements. Weight loss in the obese helps to normalise lipid levels in most conditions.

Nutrient		
Fat	< 30	Saturated fat should provide 10 per cent or less of total energy—the rest should be from mono- or poly-unsaturated fatty acids. Ratio of polyunsaturated:saturated fats should be about 0.8. More severe restrictions in particular patients and in those with lipoprotein lipase or apo CII deficiency.
Carbohydrate	50–60%	Unprocessed fibre-rich carbohydrate should predominate. Ideally 20 g fibre/ 1000 kCals. Pectins and other gel-forming fibres (e.g. those derived from various cooked beans) are particularly useful for LDL lowering.
Protein	10–15%	Vegetable proteins derived from legumes (e.g. soya) may help to decrease LDL.
Sugar		Should be severely restricted in the obese or if triglycerides are raised.
Alcohol		Should be severely restricted in the obese or if triglycerides are raised. A sensible maximum for others is 12 units per week.
Cholesterol		Less than 300 mg/day.

Practical dietary advice

If people are to be encouraged to change their diet they need to be aware of the potential benefits. They also need sufficient knowledge about the composition of various foods to make the appropriate changes.

For those who have been on an average 'Western' diet, their saturated fat intake will generally have come from four main sources. Meat and meat products will have supplied about a quarter of the total, and butter and margarine another quarter. Milk often accounts for an eighth, and milk and cheese together constitute a third for some people. Cooking fats often account for another third. Chocolate, pastries, biscuits, cakes, and convenience foods are often high in saturated fat, but because it is 'hidden' it is often ignored when an individual considers his/her fat intake. Most packaged and processed foods give little information about the food composition of the contents. Hard margarines may be labelled as vegetable fats but the hydrogenating process will have converted most of the PUFAs to saturated fat. Values for percentage fat expressed as grams of fat per 100 g food weight are often uninformative as the amount of water present is so different, between, for example, milk and sausages. Values which are most helpful in diet planning are the percentage of total energy of the food provided by fat, and these are given in Table 8.5. Some foods such as pork pies may be a quarter fat by weight and three quarters of the energy may be from fat.

In addition to a high fat intake, most people will have been consuming relatively little dietary fibre.

For many people the provision of a simple diet sheet together with recipes and advice on food preparation will be a major step in the right direction. Nothing can replace individual consultation with a dietitian, but where this is not available the diet sheet shown in Table 8.6*a*, the principles outlined in Table 8.7 and the hints and sample recipes given in the Appendix should be helpful. Many people have rather conservative ideas about food and need advice and ideas to create interesting low-fat meals. They should be encouraged to try foods which may not have been part of their previous diet, such as kidney beans, chick peas, lentils, wholemeal pasta, etc. Several food manufacturers now produce low-fat products including skimmed milk, yoghurts, soft and hard cheeses, and prepared meals. These do contain less fat than their counterparts, but

Table 8.5. Fat content of different foods, by percentage of total energy

Product	Serving	KCals	% total energy available from: Protein	Fat	Carbohydrate
Skimmed milk	⅓ pt	66	40	1	59
Whole milk	⅓ pt	130	20	52	28
Yoghurt (low-fat, flavoured)	small carton	122	23	10	67
Cream (double)	2 tbsp	179	1	97	2
Cottage cheese	100 g	96	56	38	6
Cheddar cheese	60 g	244	26	74	—
Roast chicken	120 g	178	68	32	—
Grilled white fish	100 g	95	88	12	—
Lean ham	2 thin slices	72	61	39	—
Pork sausages	2-grilled	382	16	70	14
Pork chop	250 g	645	34	66	—
Pork pie	Mini	451	10	64	26
Beefburger	Small	158	31	59	10
Chips	60 g portion	152	6	37	57
Boiled potatoes	2	96	6	—	94
Chocolate (milk)	Small bar	265	6	50	44
Biscuits	1 Digestive	71	8	38	54
Fruit cake	1 med. slice	212	5	31	64
Crisps	Small packet	133	4	60	36
Bread (wholemeal)	1 med. slice	108	15	9	76
Rice (boiled)	180 g	222	5	2	94
Cornflakes	Medium bowl	111	9	3	90
Apples	1	55	2	—	98
Vegetables (green)	100g	18	30	—	70
Baked beans	Small tin	144	31	7	62

even the 'low-fat' hard cheeses often contain a substantial quantity of fat, and intake must remain controlled. This also applies to some low-fat spreads which are lower in calories, but still contain a considerable proportion of saturated fat. Some idea of the fat content of an average helping of a number of foods, and the percentage energy from fat, protein, and carbohydrate are given in Table 8.5.

Foods in the advisable column in Table 8.6*a* and *b* are generally low in fat or high in fibre and should be used regularly. Foods 'not advised' contain large proportions of saturated fats and therefore should be avoided wherever possible. Foods 'in moderation' contain smaller amounts of fat or polyunsaturated fats or are high in 'empty' calories. Advice about the consumption of foods in these columns depends on the stage of the diet recommended. The interpretation of moderation obviously varies and more details are usually needed: for example, in a stage A diet, red meat is usually restricted to less than three times per week and medium fat cheese to twice a week. In a stage B diet, intake of these foods will probably have to be limited to once or twice a week only. Stage B diets are thus stringent, but are acceptable by some people who are very keen to avoid drug treatment.

A step-wise programme for dietary modification is given in Figure 8.1.

Most patients eating a traditional Western diet, who comply rigidly with a lipid-lowering diet, will achieve a reduction of approximately 20 per cent in their plasma cholesterol. This may be greater if weight loss is also achieved where appropriate. However there is a great deal of individual variation in response, and the precise diagnosis is also an important factor in determining response. Patients with familial hypercholesterolaemia usually respond less to diet than those with common hyperlipidaemia, or familial combined or remnant hyperlipidaemia. As a general rule drug therapy should not be considered until maximal response to dietary change has been achieved. This often takes several months. If patients who were expected to respond to diet do not respond adequately, then their diet should be reviewed and they may need to reduce the number and frequency of the foods they have been having from the 'in moderation' column.

It is important to point out that diet recommended for hyperlipidaemia is appropriate for the general population and as such is suitable for other family members who may not have the condition. It is very similar to the diet recommended for diabetes (though in those requiring insulin, all carbohydrate will need to be distributed appropriately during the day)

Table 8.6a. Stage A diet.

Eat regularly	Eat in moderation OCCASIONALLY	Eat in moderation SPECIAL TREATS	Avoid eating
Wholemeal flour, oatmeal wholemeal bread, whole grain cereals, porridge oats, crispbreads, brown rice, wholemeal pasta, cornmeal, untoasted sugar free muesli.	White bread. White flour. White rice & pasta. Water biscuits.	Sugar-coated cereals. Plain semi-sweet biscuits. Ordinary muesli.	Sweet biscuits, cream-filled biscuits, cream crackers, cheese biscuits, croissants.
All fresh, frozen, dried & unsweetened tinned fruit. All fresh, frozen, dried & tinned vegetables (especially peas, baked beans, broadbeans and lentils). Baked potatoes (eat skins).		Fruit in syrup. Crystallized fruit. Avocado. Chips & roast potatoes cooked in suitable oil.	Chips & roast potatoes Crisps & savoury snacks.
Walnuts. Chestnuts.		Peanuts & most other nuts e.g. almonds, hazelnuts, Brazilnuts	Coconut.
All fresh & frozen fish, e.g. cod, plaice, herring, mackerel. Tinned fish in brine and tomato sauce e.g. sardines & tuna.	Fish fried in suitable oil.	Prawns, lobster, crab, oysters, molluscs, winkles. Fish tinned in oil (drained)	Fried scampi.
Chicken, turkey, veal, rabbit Game. Soya protein meat substitute.	Lean beef, pork, lamb, ham & gammon. Very lean minced meat.	Liver, kidney, tripe, sweetbreads. Grilled back bacon.	Sausages, luncheon meats, corned beef, pate, salami, streaky bacon. Duck, goose, meat pies & pasties, Scotch eggs. Visible fat on meats, crackling, chicken skin.

Skimmed milk, soya milk, powdered skimmed milk. Cottage cheese. Low fat curd cheese. Low fat yoghurt. Egg white	Semi-skimmed milk.	Medium fat cheeses, e.g. Edam, Camembert, Gouda, Brie, Cheese spreads. Half fat cheeses labelled 'low fat'. Sweetened condensed skim milk.	Whole milk & cream. Full fat yoghurt. Cheese e.g. Stilton, Cheddar, Cream cheese. Evaporated or condensed milk. Imitation cream Excess eggs.
Small amounts from, next column.	Margarine & shortening *labelled "high in polyunsaturates"*. Corn oil, sunflower oil, soya oil, safflower oil, grapeseed oil. Olive oil.		All margarines, shortenings & *oils not labelled "high in polyunsaturates"*. Butter, lard, suet & dripping. Vegetable oil or margarine of unknown origin.
Jelly (low sugar) Sorbet. Fat free homemade soups. Low fat, low sugar yoghurt. Low fat natural yoghurt.	Pastry, puddings, cakes, biscuits, sauces, etc. made with wholemeal flour & fat or oil as above. Salad dressing made with suitable fat or oil as above.	Packet soups.	Pastries, puddings, cakes, & sauces made with whole milk and fat or oil as above. Suet dumplings or puddings. Salad dressing or mayonnaise made with unsuitable oil. Ice cream. Cream soups.
Marmites, Bovril, chutneys & pickles. Sugar free artificial sweeteners.	Fish & meat pastes. Peanut butter. Low sugar jams & marmalade.	Boiled sweets, fruit pastilles & jellies. Jam, marmalade, honey.	Chocolate spreads. Chocolate, toffees, fudge, butterscotch. Carob chocolate. Coconut bars
Tea, coffee, mineral water, fruit juices (unsweetened).	Alcohol.	Sweetened drinks. Squashes, fruit juice. Malted milk or hot chocolate drinks made with skimmed milk.	Whole milk drinks. Cream based liqueurs.
Herbs, spices, Tabasco, Worcestershire Sauce, Soy sauce, lemon juice.	Homemade dressings & mayonnaise made with suitable oils.	'Low fat' or 'low calorie' mayonnaises & dressings. Parmesan cheese.	Ordinary or cream dressings & mayonnaises.
EAT REGULARLY Choose from this group daily	EAT IN MODERATION —Occasionally = moderate amounts 2–3 times per week. —Special Treats = moderate amount once a week or less.		

Table 8.6b. Stricter diet for persistent hyperlipidaemia—stage B

Eat regularly	Eat in moderation	Avoid eating	
Wholemeal flour, oatmeal wholemeal bread, whole grain cereals, porridge oats, crispbreads, brown rice, wholemeal pasta, cornmeal, untoasted sugar free muesli.	White bread. White flour. White rice & pasta. [if wholemeal varieties unavailable]	Sugar-coated cereals. Plain semi-sweet biscuits. Ordinary muesli.	Sweet biscuits, cream-filled biscuits, cream crackers, cheese biscuits, croissants.
All fresh, frozen, dried & unsweetened tinned fruit. All fresh, frozen, dried & tinned vegetables (especially peas, baked beans, broadbeans and lentils). Baked potatoes (eat skins).		Fruit in syrup. Crystallized fruit. Avocado. Chips & roast potatoes	Chips & roast potatoes Crisps & savoury snacks.
Walnuts. Chestnuts.		Peanuts & most other nuts e.g. almonds, hazelnuts, Brazilnuts [N.B. some allowed if vegetarian]	Coconut.
All fresh & frozen fish, e.g. cod, plaice, herring, mackerel. Tinned fish in brine and tomato sauce e.g. sardines & tuna.	Fish fried in suitable oil. [once/week]	Prawns, lobster, crab, oysters, molluscs.	Taramasalata. Fried scampi.
Chicken, turkey, veal, rabbit Game. Soya protein meat substitute.	Lean beef, pork, lamb, ham & gammon. Very lean minced meat. [twice/week]	Liver, kidney, tripe, sweetbreads. Grilled back bacon.	Sausages, luncheon meats, corned beef, pate, salami, streaky bacon. Duck, goose, meat pies & pasties, Scotch eggs. Visible fat on meats, crackling, chicken skin.

Skimmed milk, soya milk, powdered skimmed milk. Cottage cheese. Low fat curd cheese. Low fat yoghurt. Egg white	Low fat cheese/spread cheese [once/week]	Medium fat cheeses, e.g. Edam, Camembert, Gouda, Brie, Cheese spreads.	Whole milk & cream. Full fat yoghurt. Cheese e.g. Stilton, Cheddar, Cream cheese. Evaporated or condensed milk. Imitation cream Excess eggs.
Small amounts from, next column.	Margarine & shortening *labelled "high in polyunsaturates"*. Corn oil, sunflower oil, soya oil, safflower oil, grapeseed oil. Olive oil.		All margarines, shortenings & *oils not labelled "high in polyunsaturates"*. Butter, lard, suet & dripping. Vegetable oil or margarine of unknown origin.
Jelly (low sugar) Sorbet. Fat free homemade soups. Low fat, low sugar yoghurt. Low fat natural yoghurt.	Pastry, puddings, cakes, biscuits, sauces, etc. made with wholemeal flour & fat or oil as above. Salad dressing made with suitable fat or oil as above. [small serving twice a week]	Packet soups.	Pastries, puddings, cakes, & sauces made with whole milk and fat or oil as above. Suet dumplings or puddings. Salad dressing or mayonnaise made with unsuitable oil. Ice cream. Cream soups.
Marmites, Bovril, chutneys & pickles. Sugar free artificial sweeteners. Small amounts of low sugar jam/marmalade.	Fish & meat pastes. [thin spread 2–3 times/week] Jelly.		Chocolate spreads. Chocolate, toffees, fudge, butterscotch. Carob chocolate. Coconut bars.
Tea, coffee, mineral water, fruit juices (unsweetened).	Boiled sweets Low calorie squashes	Sweetened drinks. Malted milk or hot chocolate drinks	Whole milk drinks. Cream based liqueurs.
Herbs, spices, Tabasco, Worcestershire Sauce, Soy sauce, lemon juice.	Homemade dressings & mayonnaise made with suitable oils.	'Low fat' or 'low calorie' mayonnaises & dressings. Parmesan cheese.	Ordinary or cream dressings & mayonnaises.

EAT REGULARLY—Choose from this group daily
EAT IN MODERATION—Moderate quantity once or twice a week

Table 8.7. Some practical ways to reduce saturated fat in the diet

- Substitute polyunsaturated margarine for butter.
- Use polyunsaturated oil.
- Use skimmed milk instead of full-cream milk. Avoid cream.
- Replace ordinary hard cheeses with low-fat or reduced fat varieties, e.g. cottage cheese, quark (skimmed milk soft cheese), Tendale, Shape, or supermarket's own brands of half-fat cheese.
- Try low-fat yoghurt and skimmed milk cheeses in place of cream, mayonnaise and salad cream, or use a small quantity of olive oil.
- Choose lean cuts of meat and trim fat off before cooking.
- Eat smaller portions of meat. Extend meat and poultry dishes by using pulses, cereals, and vegetables.
- Eat chicken, turkey and fish more often as these are lower in fat. Remove all skin and fat from poultry.
- Grill, steam, poach, bake, braise, or casserole instead of frying or roasting with extra fat. Do not use lard.
  To exclude the need to add fat when roasting, wrap food in foil or roasting bags to retain juices and prevent drying out. Meat can be roasted without fat, or cooked on a rack over a pan of hot water.
- Low fat sauces can be made by mixing flour or cornflour with cold water, skimmed milk, or stock before cooking to thicken.
- Skim off fat from stews and soups by removing it with absorbent kitchen paper or, if the dish is allowed to cool first, the fat could be scooped off with a spoon.
- Be on the look-out for new low-fat or reduced-fat products in the shops. Check labels for fat content and type of fat used in processed foods.
- In addition, eat more fibre-rich foods.

and for many gastrointestinal diseases. It is only rarely in conflict with other sets of dietary recommendations.

The cost of a different diet is often a question raised by patients, particularly those on low incomes. It is possible to plan a low-fat diet which is no more expensive than a person's usual diet. Vegetables and pulses are relatively cheap, brown bread is not significantly more

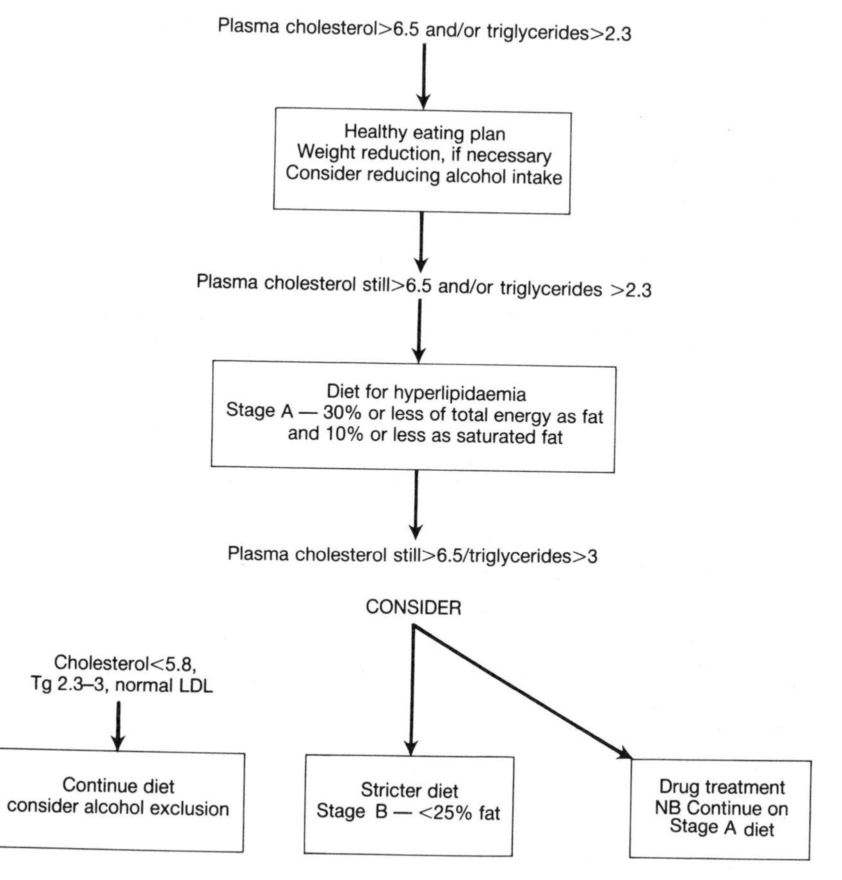

NB: This step-wise plan is suitable for some patients. For others, particularly those with marked hyperlipidaemia, the stage A (or even in a few cases the stage B) diet will need to be prescribed at the outset.

Fig. 8.1. Plan for step-wise dietary management of hyperlipidaemia.

expensive than white, fish can be cheaper than meat, and polyunsaturated margarines are less expensive than butter.

Children with hyperlipidaemia need special consideration. In some instances familial hypercholesterolaemia is diagnosed very early in life. Very young children should not be given a very low-fat diet, but from the age of 3 or 4 years general healthy eating is encouraged, with the avoidance of foods containing excessive saturated fat. Dietary recommendations for older children need to take into account the overall nutritional requirements for growth but this is easily achieved within the basic context of a diet low in saturated fat and high in soluble fibre.

For many people with hyperlipidaemia dietary change is initially difficult, but with appropriate teaching the majority realise that the change not only produces a reduction in blood lipids but provides a palatable alternative to the traditional high fat 'Western diet'.

Summary

People with hyperlipidaemia should alter their calorie intake to enable them to attain their ideal body weight. Saturated fat intake needs to be reduced. This is best achieved by substituting polyunsaturated fat where suitable and restricting fat in as many contexts as possible. Unrefined carbohydrate should replace refined carbohydrate containing food where possible, and sugar and alcohol intake may need to be reduced. Consumption of vegetables, fruit, pulses, and fish should be increased.

9 Drug treatment of hyperlipidaemia

Many patients with a familial hyperlipidaemia require treatment with drugs in addition to dietary control, as do some patients without an obvious family history. Drug treatment will depend on age, medical history, and family history, as well as other factors. However, it should not be commenced until there has been a careful trial of the most rigorous diet appropriate for the particular patient. In addition, even if drug treatment is required, patients must be advised to continue on an appropriate diet.

A number of drugs are currently available for the treatment of hyperlipidaemia (Table 9.1). Some are most suitable for the treatment of raised cholesterol, others are used to lower both cholesterol and triglycerides, or substantially raised triglycerides.

Cholestyramine and colestipol

Cholestyramine and colestipol are a first-line treatment for hypercholesterolaemia caused by a high LDL cholesterol. They are anion exchange resins which are not absorbed into the body but have their main action in the intestine. The resins complex with bile salts in the intestine and result in their excretion in the faeces, rather than re-absorption and use. This loss means that more bile salts need to be synthesized, and this involves the use of cholesterol (Fig. 9.1). Depletion of cell cholesterol leads to an increase in LDL receptors and LDL removal from the plasma. Constant use of these resins thus results in a lowering of the plasma cholesterol. Total and LDL cholesterol are often reduced by 15–25 per cent, and HDL cholesterol may rise slightly. Triglycerides may, however, rise slightly so these drugs should not be used alone in patients who have a plasma triglyceride above 3 mmol/l, and the effect on plasma triglycerides should be monitored in anyone with a level exceeding the reference range. The daily dosage of cholestyramine required varies from 4 to 28 g. The resin is packaged in 4 g sachets and should be dissolved in an appropriate quantity of water and shaken to give a good suspension prior to ingestion. A marked glass is available to

Table 9.1. Drug therapy

Drug		
Cholestyramine (Questran)		
	Availability	—4 g sachets
	Dosage	—1–8 sachets/day
Colestipol (Colestid)		
	Availability	—5 g sachets
	Dosage	—1–6 sachets/day
Probucol (Lurselle)		
	Availability	—250 mg tablets
	Dosage	—500 mg twice a day
Bezafibrate (Bezalip)		
	Availability	—200 mg tablets
		—400 mg long acting preparation (Bezalip mono)
	Dosage	—200 mg three times a day or 400 mg Bezalip mono/day
Clofibrate (Atromid-S)		
	Availability	—500 mg capsules
	Dosage	500 mg twice or three times a day
Gemfibrozil (Lopid)		
	Availability	—300 mg tablets.
	Dosage	—300 mg three or four times a day
Nicotinic acid derivates (see text)		

help assess quantity. The mixture is slightly unpalatable, but the addition of fruit juice usually improves its acceptability. Some people find it convenient and preferable to leave the suspension in the refrigerator for several hours and decant a quantity at appropriate times. Other ideas are given in the Appendix for those who dislike taking it in this way.

It is wise to start patients on a low dose—one or two sachets a day—and gradually increase if necessary, because flatulence and mild abdominal discomfort sometimes occur. If constipation results then this may be eased by an increase in dietary fibre. Malabsorption is very rare, but when the drug is used in children a folic acid preparation is given. The resins may, however, interfere with the absorption of other drugs, and it is a sensible precaution to take other medication at a different

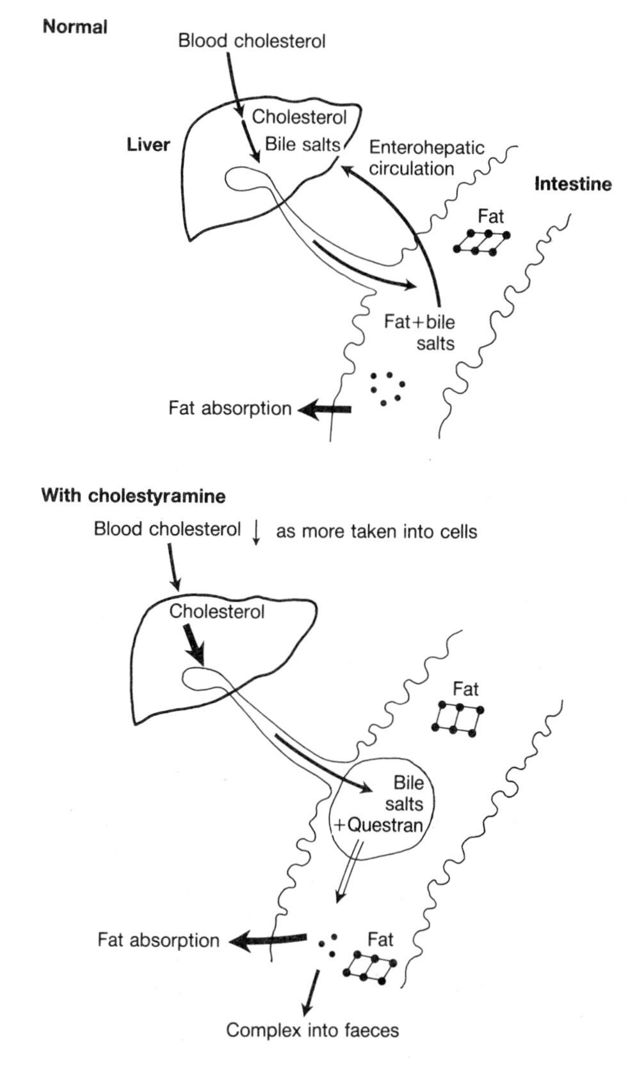

Fig. 9.1. Mechanism of action of cholestyramine.

time. A few patients, particularly those with pre-existing gastrointestinal problems, are unable to tolerate the dose of cholestyramine required. It is then sensible to try another drug, or a combination of medications with only a small dose of resin.

Probucol

In the recommended dose of 500 mg b.d. probucol gives a variable response, but it may lower the plasma LDL cholesterol by 10–15 per cent when used in conjunction with an appropriate diet. Its maximal effect occurs after 1–3 months treatment, but it appears to be less effective than cholestyramine and it has the disadvantage of reducing HDL cholesterol levels as well as LDL. The specific mode of action is unclear, although it probably enhances removal of LDL from the circulation, and it does usually reduce the size of xanthomas. Oral absorption is low, but appears to be more uniform when the drug is taken with food. Accumulation in adipose tissue occurs, with the result that the drug may be present in the body for months after the last dose.

Tolerance of probucol is good, with only occasional patients reporting nausea and flatulence. It can, however, cause prolongation of the QT interval and could potentiate arrythmias, so it should be used cautiously in patients with ischaemic heart disease. The safety of the drug in children or during pregnancy has not been established.

Fibric acid derivatives

There are a number of fibrates licenced for use in different countries. They include *clofibrate, bezafibrate, gemfibrozil*, and *fenofibrate*. They act by several mechanisms, and their main effect is to reduce VLDL cholesterol. This causes a fall in plasma cholesterol and triglyceride levels.

Clofibrate

The WHO clofibrate trial showed that the treatment of hyperlipidaemic men with this drug reduced their morbidity from myocardial infarction. There was, however, an increase in deaths from other causes and in the incidence of gallstones. Owing to this it is not commonly recommended, although in remnant hyperlipidaemia it may have a dramatic effect, better than that of other fibrates, and in this context the benefit is likely to outweigh any risks.

Bezafibrate

This drug is an analogue of clofibrate and has been a first-line drug in the treatment of familial combined hyperlipidaemia. It is also of value in patients with familial hypercholesterolaemia—both in those who are unable to tolerate cholestyramine, and as an additional agent in those with inadequate response. Bezafibrate is given as 200 mg t.d.s. or as a once daily 400 mg preparation, Bezalip-mono. It reduces total cholesterol, particularly LDL cholesterol, decreases the LDL/HDL cholesterol ratio, and also reduces plasma triglycerides.

Gemfibrozil

This drug is another analogue which has similar properties. It was first marketed in the U.K. in 1986, but it has been used in the U.S.A. for many years. The large Helsinki Heart trial has shown the beneficial effects of lipid lowering using this drug (Chapter 4). There was a 34 per cent reduction in coronary events associated with mean falls in cholesterol, LDL cholesterol, and triglycerides of 8 per cent, 9 per cent, and 40 per cent, respectively. The incidence of side-effects was not appreciably different from placebo. The drug is, however, considerably more expensive than bezafibrate and is not currently available as a single daily dose.

Side-effects

The incidence of side-effects from bezafibrate and gemfibrozil is low; those most frequently experienced are gastrointestinal, such as nausea and a feeling of fullness. Impotence may sometimes occur, although this seems more common on bezafibrate than gemfibrozil. Myositis, such as was occasionally seen with clofibrate, occurs very rarely. The drugs are contraindicated during pregnancy and in people with severe renal or hepatic disease. There is little information available about their safety and effectiveness in nephrotic syndrome, but if they are used then the dose must usually be reduced. The fibrates may potentiate the action of warfarin and patients requiring therapy with these two drugs must be very carefully monitored. Unlike clofibrate, bezafibrate and gemfibrozil have little effect on the lithogenic index of bile, and are probably not associated with a marked increase in the incidence of gallstones.

In some people with combined hyperlipidaemia the use of fibrates may increase LDL cholesterol, although reducing total cholesterol and triglyceride. If this occurs then a small dose of a bile acid sequestrant can be added.

Thyroxine

The discovery that the treatment of patients with hypothyroidism caused a reduction in their plasma cholesterol level led to attempts to treat euthyroid patients with thyroid hormones. Thyroid hormones do affect cholesterol metabolism. They increased cholesterol synthesis in the liver, but also increase its faecal excretion and conversion to bile salts. Clearance of LDL from the circulation is increased. In the 1950s, D-thyroxine was found to lower the plasma cholesterol concentration without causing the increase in metabolic rate which L-thyroxine did. D-thyroxine can reduce plasma LDL by up to 15 per cent, but does not alter HDL, and it can increase the frequency of anginal attacks in patients with CHD. In the Coronary Drug Project, in the early 1970s (see Chapter 4), survivors from a myocardial infarction who were given D-thyroxine had a decreased plasma cholesterol, but an increased mortality from coronary disease. Some patients also developed abnormal liver function tests.

This drug is occasionally used in young people without coronary disease who show a significant cholesterol reduction and are unable to tolerate any other medication.

Nicotinic acid derivatives

These preparations reduce VLDL and LDL synthesis and thus cause a fall in LDL cholesterol and triglyceride levels. The effect tends to be variable when the medication is used alone, but in some patients, particularly those with familial combined hyperlipidaemia, they may provide an additional benefit when a single drug has insufficient effect.

Patients often experience flushing and a sensation of warmth after taking these preparations, and treatment is best started with a low dose, given several times a day with food, and increased gradually. The flushing is mediated by prostaglandins and can be reduced by low doses of aspirin. Nicotinic acid itself often has unacceptable side-effects, but some patients may tolerate analogues such as nicofuranose, nicangin (available from hospital pharmacy only), acipimox or inositol nicotinate, although these may give less lipid lowering. Gout and abnormalities of liver function may sometimes occur.

Maxepa

The omega 3 fatty acids in fish oil may be of use in some patients with severe hypertriglyceridaemia, but more data is needed, particularly in comparison with other treatments. Use for hypercholesterolaemia is not indicated by the present Committee of Safety of Medicines licence. The effect on LDL cholesterol seems to depend on the dose, and at low doses LDL may actually rise, although VLDL falls. The value of any changes it causes in the clotting mechanism and its interaction with any other drugs remain to be shown. Antioxidants have to be added to these very unsaturated fatty acids to avoid the formation of toxic oxidation products.

HMGCoA reductase inhibitors

A promising new approach to the treatment of patients with hypercholesterolaemia refractory to conventional treatment, is the development of drugs which act on a different part of cholesterol metabolism to those already described. These drugs are chemicals which act within the cells to block the action of HMGCoA reductase, the enzyme responsible for the synthesis of cholesterol. In practice, when the cholesterol supply to the cell from the plasma is reduced by drugs such as cholestyramine, the endogenous synthesis of cholesterol may increase. This resets the balance somewhat with the net result that plasma cholesterol does not fall to the extent expected. By blocking the enzyme responsible for cholesterol synthesis a greater lipid-lowering effect can be achieved. The use of such drugs, which include lovastatin, simvastatin, and SQ 31 000, is being investigated in a number of centres. Licences have been granted in some countries. The preliminary results appear promising, with reductions in LDL cholesterol of up to 45 per cent. Abnormalities of liver function tests sometimes occur and this needs further evaluation. The plasma creatine kinase level may also rise transiently. Ophthalmic examination is currently necessary to check that lens opacities do not occur. HMGCoA reductase inhibitors are potentially teratogenic so their use will generally have to be confined to men and women who are post-menopausal or surgically sterile.

Figure 9.2 shows a plan for the treatment of a person with hyperlipidaemia.

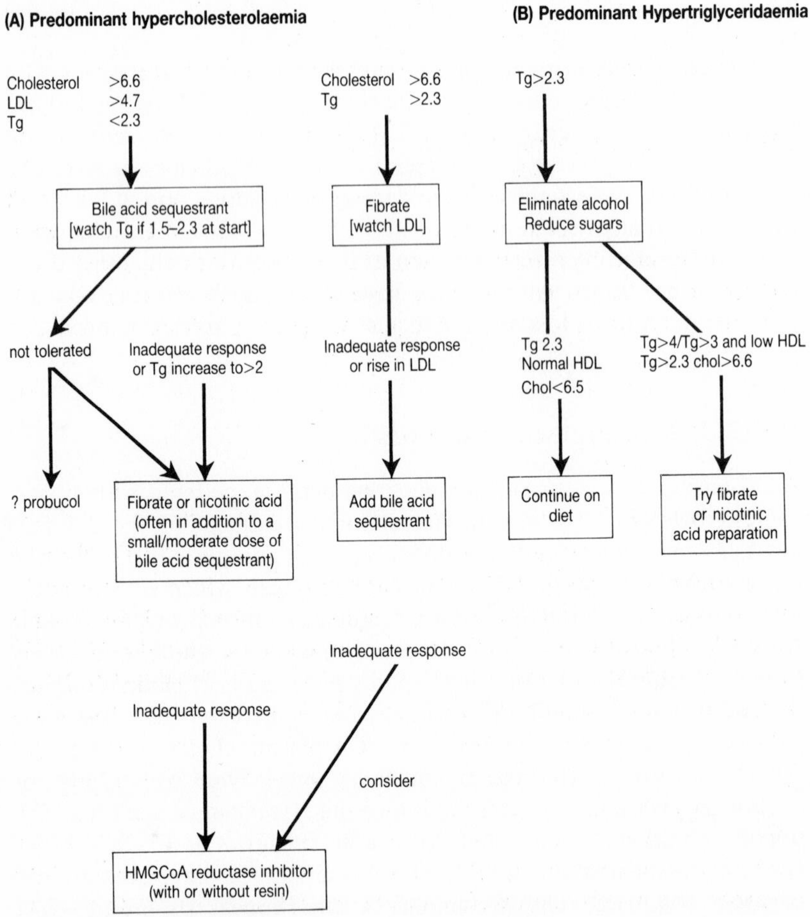

(A) Predominant hypercholesterolaemia

Cholesterol >6.6
LDL >4.7
Tg <2.3

Bile acid sequestrant
[watch Tg if 1.5–2.3 at start]

not tolerated

? probucol

Inadequate response
or Tg increase to>2

Fibrate or nicotinic acid
(often in addition to a
small/moderate dose of
bile acid sequestrant)

Cholesterol >6.6
Tg >2.3

Fibrate
[watch LDL]

Inadequate response
or rise in LDL

Add bile acid
sequestrant

(B) Predominant Hypertriglyceridaemia

Tg>2.3

Eliminate alcohol
Reduce sugars

Tg 2.3
Normal HDL
Chol<6.5

Continue on
diet

Tg>4/Tg>3 and low HDL
Tg>2.3 chol>6.6

Try fibrate
or nicotinic
acid preparation

Inadequate response

Inadequate response

consider

HMGCoA reductase inhibitor
(with or without resin)

Fig. 9.2. Flow diagram of the suggested treatment of primary hyperlipidaemia in a young/middle-aged adult.

Many patients with hyperlipidaemia have a very satisfactory decrease in plasma cholesterol when undergoing dietary control in combination with single drug treatment. Some patients do, however, require additional drug treatment. The reasons for long-term treatment and its importance must be emphasized to the patients, as compliance may be reduced when they do not feel any immediate benefit.

Patients with *homozygous familial hypercholesterolaemia* seldom achieve acceptable cholesterol levels, even with maximal drug therapy. Additional measures may then be considered in these patients.

Plasma exchange/plasmaphoresis

This form of treatment is sometimes used in patients with very high cholesterol levels, such as those with homozygous familial hypercholesterolaemia. The treatment temporarily reduces the cholesterol, but the procedure has to be repeated every few weeks. Plasma replacement also has potential risks. Several types of specific aphoresis are being developed which specifically remove LDL and VLDL and do not require plasma infusion, but these will be expensive and will require evaluation.

Surgical procedures

Ileal bypass has been performed in a few patients with severe hypercholesterolaemia resistant to drug treatment, and in some patients with homozygous familial hypercholesterolaemia. The result of this surgical procedure is a reduction in the reabsorption of bile acids, which usually occurs as part of the enterohepatic circulation. Unfortunately the operation frequently has unwanted side-effects, including severe diarrhoea and metabolic disturbances, and is rarely performed in the U.K.

Summary

Currently, cholestyramine is generally regarded as a first-line drug treatment for hypercholesterolaemia due to high LDL cholesterol levels, although the HMGCoA reductase inhibitors are likely to prove very valuable for some people, and the fibrates are also used.

Bezafibrate or gemfibrozil are first-line drug treatments for patients with raised cholesterol and triglyceride levels, and for those with significant primary hypertriglyceridaemia. Addition of cholestyramine may sometimes be needed if the LDL rises. A small number of patients will need a combination of two drug types.

10 Overview of the familial hyperlipidaemias and their treatment: other aspects of management

Familial hypercholesterolaemia

Incidence	Approximately 1 in 500 in Britain
Inheritance	Autosomal dominant
Cause	Genetic defect. Usually reduced number of LDL receptors
Effect	High plasma LDL cholesterol
Clinical signs	
	Tendon xanthomas
	Xanthelasmas
	Early corneal arcus
Treatment	
(a) Diet	Low saturated fat diet, increase in P:S ratio, high fibre, appropriate calories to attain ideal body weight
(b) Drugs	Bile acid sequestrant, for example, cholestyramine or fibrate, or HMGCoA reductase inhibitors (when appropriate)

(c) Reduce other CHD risk factors
Follow-up as appropriate
Screen family

Familial combined hyperlipidaemia

Incidence	Approximately 1 in 300 in Britain
Inheritance	Likely to be autosomal dominant
Effect	High plasma VLDL cholesterol (cholesterol and triglyceride raised)
	High risk premature CHD

Treatment

(a) Diet — Low saturated fat diet, increase in P:S ratio, high fibre, appropriate calories to attain ideal body weight

(b) Drugs — Fibric acid derivative, perhaps with a small dose of a bile acid sequestrant if LDL cholesterol raised

(c) Reduce other CHD risk factors

Follow-up as appropriate

Screen family members >18 years. Tell parents about reminding younger children to be tested at 18 and 25 years

Remnant hyperlipidaemia

Incidence	1 in 10 000
Inheritance	?
Cause	Apo E2/E2 plus another primary or secondary factor
Effect	High risk CHD High risk peripheral vascular disease
Clinical signs	Tubo-eruptive xanthomas

Treatment

(a) Identification of any secondary factor and treatment

(b) Diet

(c) Drug treatment with a fibrate if necessary

Familial hypertriglyceridaemia

Incidence	Approximately 1 in 1000
Inheritance	Probably autosomal dominant
Clinical signs	Eruptive xanthomas if triglyceride concentrations markedly raised
Effect	Risk of pancreatitis if concentrations exceed 20 mmol/l

Treatment

(a) Secondary factors such as diet, weight, glucose intolerance, and alcohol intake are often very important and treatment of these factors may reduce levels significantly

(b) Drug treatment—if levels remain elevated (generally >4 mmol/l), especially if the HDL cholesterol is low. Fibrates are often effective

Genetic counselling

If one parent has familial hypercholesterolaemia there is a 50 per cent chance that any children will inherit the gene and have the condition. The risk is probably the same in familial combined hyperlipidaemia.

Heterozygous familial hypercholesterolaemia can often be diagnosed at birth, although testing is often postponed until the child is older. The wishes of the parents as to the timing of the diagnosis in children should be considered. Testing between the ages of 4 and 6 is probably sensible. Interpretation of the results should be performed by someone with experience in diagnosis in children as the reference range is very different in young children (Chapter 6). Even then it may not be possible to confirm or exclude the condition in all children. A high level can be confirmative, but a borderline or high 'normal range' level may be found, particularly when the family are on a low-fat diet. In these cases repeat testing should be arranged a year or two later.

In families with familial combined hyperlipidaemia testing should not be performed until the offspring are in their late teens or early 20s, as the lipid are seldom raised in childhood. Even at this age the disorder may not be manifest or the levels may only be marginally elevated. Follow-up and repeat testing should therefore be offered in the late 20s.

The revelation that there is a serious genetic disorder in the family can be absolutely devastating, even for unaffected members. A parent with premature cardiovascular disease needs reassurance that with effective treatment the prognosis for their child is likely to be much better than their own. Feelings of guilt that they have passed on illness on to their children may remain, however. In the unlikely case of two individuals with familial hypercholesterolaemia marrying then there is a 1 in 2 chance that any child will inherit one gene and have heterozygous familial hypercholesterolaemia and a 1 in 4 chance that they will inherit two genes resulting in the very severe homozygous form.

Treatment of other risk factors

The importance of risk factors other than hyperlipidaemia were discussed in Chapter 3.

Smoking

Cigarette smoking causes an increased incidence of CHD, with a relatively greater risk in people under the age of 50. It acts synergistically with risk factors such as hypercholesterolaemia. There is evidence that smoking may adversely affect lipid levels by slightly reducing the plasma HDL and the HDL/LDL ratio. Surveys show that a large percentage of smokers would like to give up and feel that advice from their doctors would provide an incentive. All patients who smoke should ideally receive personal counselling including information on the hazards, the potential benefits of discontinuing, advice on ways to stop and to cope with the problems, and a target. Fear of a small weight gain is not a justifiable excuse for continuing to smoke as in most cases a 5–10 kg weight gain is a lesser health risk than smoking 20 cigarettes a day. Nicotine chewing gum, acupuncture, hypnosis, and group therapy can be helpful.

Some patients find it difficult to stop smoking completely while changing their diet, particularly if this involves calorie reduction. Encouraging patients to realize which cigarettes are smoked at 'habit times' and to initially stop these while implementing the diet may prove more acceptable, and improve compliance.

Hypertension

The combination of hypertension and hyperlipidaemia is relatively common. In a very few individuals treatment with thiazides or beta blockers may cause an increase in plasma triglycerides, so it is probably prudent to check the plasma lipids before starting anti-hypertensive treatment, if possible.

- *Thiazides.* Thiazide diuretics tend to increase the production of triglyceride-rich VLDL from the liver. This may cause an increase in plasma triglyceride levels in some patients, but the rise is usually clinically insignificant. In a very few patients, however, there is a marked rise in plasma triglycerides to a level which could constitute a risk for vascular problems or pancreatitis.
- *β-blockers.* The effect of different β-blockers appears to be variable, although the trend is probably similar. Several studies have shown that plasma triglycerides are increased slightly on long-term treatment; the greatest rise occurring with propranolol. Propranolol also reduces HDL cholesterol and increases the total cholesterol/HDL cholesterol ratio. Other β-blockers are therefore preferable to propranolol in patients with hyperlipidaemia. The effect of the selective β-blockers is small and it is usual to leave well-controlled hyperlipidaemic patients on them.

• *Prazosin.* Total cholesterol and triglyceride concentrations appear to be unaffected or to fall on this drug. HDL may increase.
• *Calcium antagonists.* Drugs such as nifedepine do not increase plasma lipids.
• *Nitrates and angiotensin converting enzyme inhibitors.* We are not aware of any adverse effect of these drugs on plasma lipids.

Now that more drugs are available for the treatment of hypertension, it would seem sensible, in the future, to try drugs without a potential adverse effect on plasma lipids in patients with hyperlipidaemia.

Lack of exercise

An increase in activity, particularly walking and swimming, is likely to benefit most people, especially those with a sedentary lifestyle. It can assist weight reduction and it may, independently, cause a small increase in HDL cholesterol, particularly HDL_2. Exercise may also be of psychological benefit. The degree of exertion should be increased gradually, particularly in those who already have CHD, and the exercise should be performed regularly.

Obesity

This may cause a number of medical problems which include glucose intolerance and hyperlipidaemia. Some patients who are only moderately overweight (10 per cent) may show a marked reduction in plasma lipids on weight reduction. Figure 10.1 gives an idea of the range of desirable weight by height, and a more detailed guide to ideal body weights is given in Table 8.1. The body mass index (BMI) is used to assess obesity. This index is calculated as the person's weight in kilograms divided by their $height^2$ in metres. A BMI greater than 25 indicates that someone is overweight, and an index of 30 that they are seriously obese.

Follow-up of patients trying to lose weight is important, and many overweight individuals find that joining a slimming club provides encouragement and support.

Oral contraceptives

Much has been written about the effects of various oral contraceptive preparations on plasma lipid levels. In general, oestrogens tend to increase plasma triglyceride levels and reduce plasma cholesterol. Progestogens tend to lower HDL cholesterol levels, although the effect does depend on the type of progestogen: progestogens with androgenic/

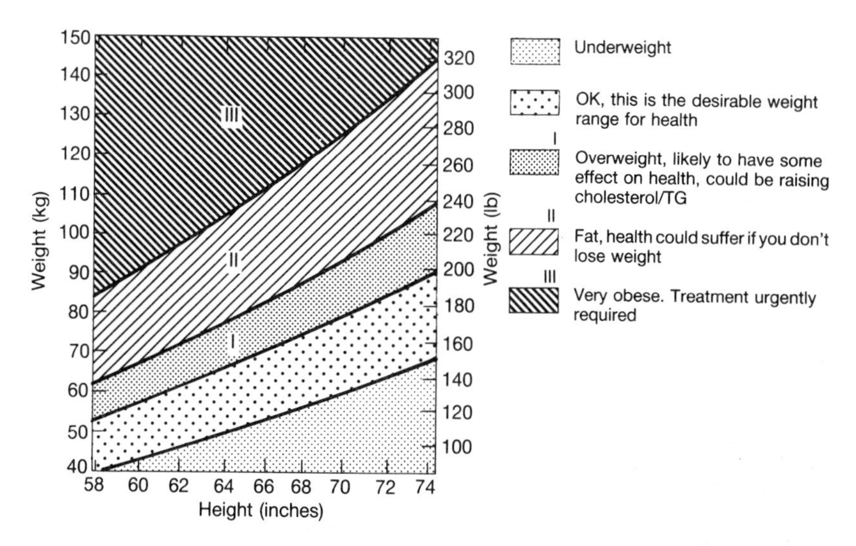

Fig. 10.1. Ideal weight. Being overweight tends to raise blood cholesterol as well as causing other problems. Therefore it is essential to achieve and maintain an ideal weight. This depends on sex, build, etc. (From Garrow, J. S. (1981). Treat obesity seriously. Churchill Livingstone, Edinburgh.)

antioestrogenic activity, such as levo-norgestrel appear to have the greatest effect. The overall effect on plasma lipids depends on the relative dose of the constituents.

Long-term use of combined oral contraceptive preparations is probably unwise in patients with hyperlipidaemia. The advisability of short-term usage depends on several factors and must be assessed for each woman. Low dose, triphasic preparations or a preparation containing a progestogen with weak androgenic activity, such as Marvelon, are likely to be most suitable for those who wish to use this method of contraception. Weak androgenic progestogen-only pills can also be considered.

The use of post-menopausal hormone replacement must also be considered carefully. Once again the choice of preparation is important, and some do have a potentially beneficial effect on plasma lipids in normolipidaemic women, and reduce morbidity. We are not, however, aware of any studies on hyperlipidaemic women and the effects may be different in those with a disorder of lipid metabolism. For this reason there is generally a reluctance to use them in hyperlipidaemic women unless there is a pressing need, but this may change as more data appears.

Effects of some other medications

Patients with CHD may be prescribed a number of different types of medication both for their CHD and for intercurrent diseases. The possible effects of any drugs on plasma lipids and any interaction with lipid-lowering drugs should be considered. For example, several case reports have indicated that amiodarone can increase plasma lipids.

Cholestyramine may interfere with the absorption of other drugs administered at the same time, and it is preferable for them to be given at least an hour apart. Cholestyramine and the fibrates may potentiate the action of warfarin, so careful monitoring of the prothrombin time is needed in people on warfarin and the dose may sometimes need to be reduced.

Psychology

When an individual is found to have hyperlipidaemia it is often a great shock to them. Suddenly they are advised to make alterations which may completely change their lifestyle. This often seems a very difficult task, particularly if they feel well and the raised lipids have been found during screening. Obviously each person is an individual in his/her knowledge, perception, needs, and aims. Motivation to make changes varies, although it is often high in those who have seen relatives die young from CHD. Intrinsic motivation is more powerful than extrinsic motivation so patients need to be allowed to feel that they are 'in control' and have responsibility for their own health. Motivation in many people increases with success, and for these people it is important that the lifestyle changes needed are structured into steps which are achievable and ordered logically. The individual's previous health beliefs need to be considered as they may affect or interfere with the acquisition and application of new knowledge. One aspect of this is their interpretation of previous information which they have been given about their condition. Checking this and correcting any misconceptions may be time well spent. Incorrect beliefs about health matters which have been held for a long time or which are the opinion of respected friends may be very difficult to modify. Encouraging people to ask questions and to work out arguments and plans for themselves, although time consuming, is likely to produce positive long-term results. The information and explanations provided often have more impact and are more readily remembered if the person feels they are directly relevant to them, rather than being rather general facts. Relatives of individuals with hyperlipidaemia sometimes ignore suggestions from affected relatives that doctors have advised blood testing. This may be because of fear of a positive result. The point has to be stressed that if they are tested then they can be offered appropriate treatment to reduce the risks.

Important aspects of the patient/health professional interview thus include:

1. Eliciting the patient's health beliefs and what they have already been told about their condition and their interpretation of this.
2. Countering misinformation and inappropriate negative attitudes, and reinforcing positive attitudes. Giving the patient more information.
3. Planning with the patient an appropriate course of action, taking into consideration the individual's circumstances.
4. Eliciting degree of motivation and how to reinforce positive changes.
5. Providing appropriate follow-up.

Section III

Screening

11 Population and high-risk strategies in coronary heart disease prevention

Much has been written about the advantages and disadvantages of the two risk reduction strategies for the prevention of premature coronary heart disease. One is aimed at the entire population and the other is targeted at high-risk individuals. Both are essential in any serious attempt to reduce premature CHD in high-risk populations. This chapter describes the two approaches and emphasizes how they are linked.

Population strategy

The population strategy is based on the recognition that where CHD is common, the majority of cases occur among those with moderately elevated levels of cholesterol and other risk factors. Figure 11.1 shows how an individual's risk increases with increasing levels of serum cholesterol, but that most of the CHD cases attributable to the cholesterol-associated risk do not occur from the few at high risk but from the large numbers exposed to a lesser risk. The population strategy aims to improve nutritional habits (and other health-oriented behaviour) so that the average level of cholesterol in the population falls. As there is no important threshold in the relation between cholesterol level and CHD (see Chapter 2) this will result in a population at less risk of CHD. If this is carried out in conjunction with a reduction in other CHD risk factors, a substantial reduction in the epidemic of premature CHD might be expected.

The principles of dietary change which might be recommended for the whole population are similar to the initial guidelines for individuals with hyperlipidaemia described in Chapter 8, although they are usually rather less restrictive.

These are, basically:

(1) control of overweight;

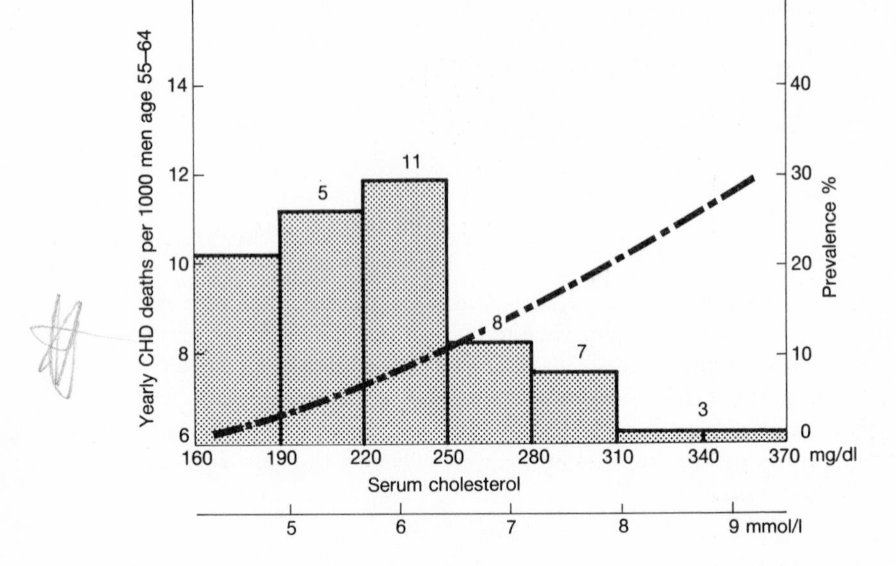

Fig. 11.1. Prevalence distribution of serum cholesterol and CHD mortality risk (—·—·—·). (From WHO Technical Report 670, 1982.) Numbers over column represent attributable deaths/1000 men in each group, per 10 years.

(2) reduction of saturated fatty acids to 10 per cent or less of total energy;
(3) increased consumption of soluble fibre;
(4) partial replacement of saturated fatty acids by mono- and poly-unsaturated fatty acids;
(5) restriction of dietary cholesterol to 300 mg, or less, per day.

The difference between the average recommendation and the current U.K. diet is given in Figure 11.2.

The nutritional guidelines issued by various national and international organizations differ both in emphasis and in the quantities recommended. For example, recommendations concerning the percentage of total energy to be derived from fat range from 25 to 35 per cent. The official British recommendations (those of the Committee on Medical Aspects of Food Policy—COMA) emphasized the need for individuals to reduce total fat to 35 per cent, or less, of their total energy. These recommendations are not as stringent as other recommendations since it was felt that the British public would not be prepared to go any further at

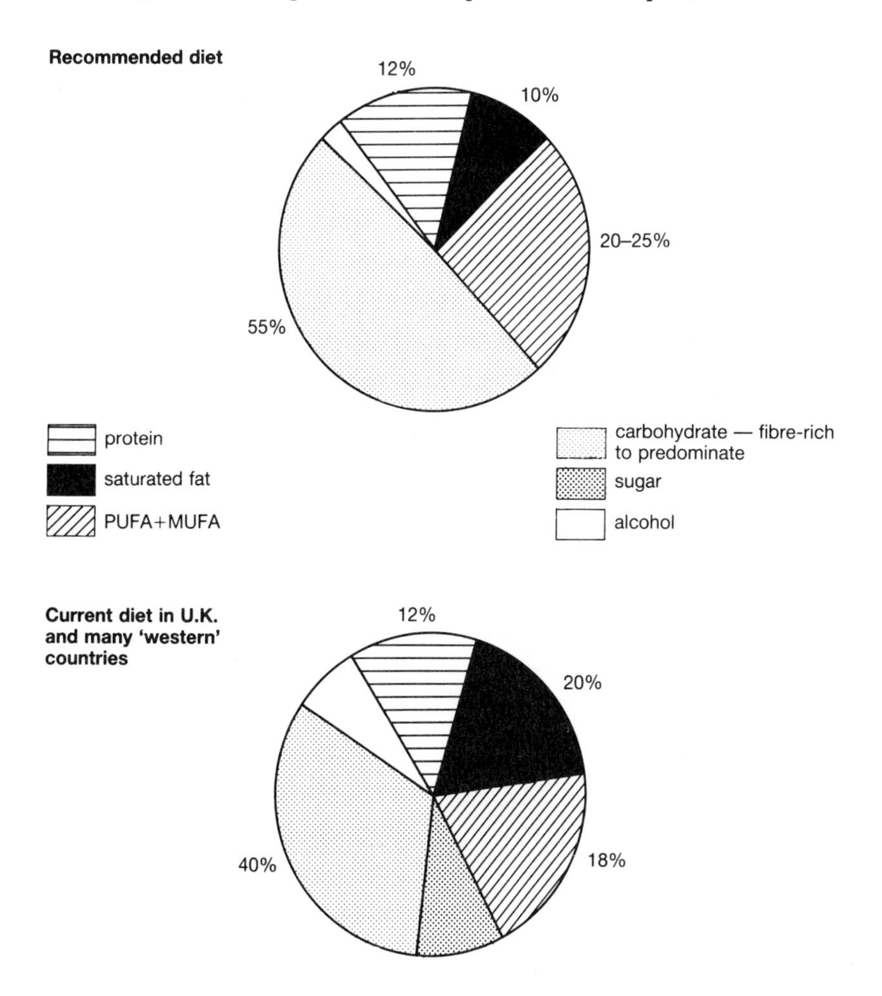

Fig. 11.2. Comparison of current UK diet with diet recommended for the population.

present. However, the British National Advisory Council of Nutrition Education (NACNE) suggested that fat be reduced so that the total intake provides less than 30 per cent of the total energy. From various published studies it is possible to predict the effect on the mean serum cholesterol levels of the population with full compliance to different sets of recommendations. Table 11.1 shows how the mean cholesterol levels

Table 11.1. Predicted effects of isocaloric dietary change on mean plasma cholesterol in men and women aged 25–59

Dietary recommendation:	Present diet	Modest change		More marked change (especially saturated fat reduction		
		COMA	NACNE	WHO	AHA	Multi-factorial
Cholesterol (mmol/l)	5.9	5.2	5.0	4.9	4.8	4.6
% reduction		12%	15%	17%	19%	23%

COMA —Committee on Medical Aspects of Food Policy
NACNE—British National Advisory Committee of National Education
WHO —World Health Organization
AHA —American Heart Association

found in 25–59-year-old men and women in the National Lipid Screening Project would be altered by the two sets of British recommendations (from COMA and NACNE), the advice of the World Health Organization (WHO), the American Heart Association (AHA), and multifactorial dietary advice which includes all five factors listed above. It is clear that full compliance with all sets of dietary guidelines would lead to considerable reduction in mean cholesterol levels.

Table 11.2 shows the percentage of people who currently have cholesterol levels exceeding certain selected cut-off points, and the effect of various reductions in cholesterol. A mean reduction of 17 per cent, predicted to result from full compliance with the WHO dietary recommendations, would leave an expected 20 per cent of the population with a plasma cholesterol greater than 5.2 mmol/l (a level associated with some degree of risk) and an expected 3 per cent with hyperlipidaemia (level > 6.5 mmol/l). Adherence to the more conservative COMA recommendations would lead to a fall of 12 per cent in serum cholesterol, but 38 per cent of the population would continue to have undesirable levels and 6 per cent would have high levels. In the more likely situation of incomplete dietary compliance the still greater prevalence of hyperlipidaemia is shown. All these calculations are made on the assumption that dietary change is isocaloric. Moderate success in reducing the

Table 11.2. Percentage of U.K. population aged 25–59 with plasma cholesterol above certain limits and predicted effect of adopting moderate dietary recommendations

Plasma cholesterol (mmol/l)	>6.5	>5.7	>5.2
Prevelance study 1984/5	23	45	63
Mean reduction of 7–9% +	12	30	50
Mean reduction of 12%*	6	23	38
Mean reduction of 17%§	3	12	20

+ incomplete compliance to COMA recommendations
* complete compliance to COMA recommendations
§ complete compliance to WHO recommendations

prevalence and extent of obesity would substantially increase the fall in cholesterol, but it is not possible to estimate by precisely how much.

Individual or high-risk strategy

It is clear from the above discussion that even with a high level of compliance to appropriate dietary recommendations, a substantial proportion of the population will continue to be at high risk of CHD because of high levels of cholesterol. Of particular importance is the fact that virtually all individuals with familial hypercholesterolaemia, the majority with familial combined hyperlipidaemia and remnant hyperlipoproteinaemia, and some people with other forms of hyperlipidaemia, do not show an adequate response, even with full compliance to the kind of dietary recommendations suitable for the general public. For them a more stringent diet (see Chapter 8) and often drug therapy are necessary, and appropriate follow-up is essential. A high-risk strategy is thus required to complement the population strategy so that these people can be identified and given individual care.

It has been suggested that selective screening of those with various 'risk attributes' for blood lipids will enable the majority of people with marked hyperlipidaemia to be detected. Tables 11.3 and 11.4 show the proportions of those with cholesterols greater than 6.5 or 8.0 mmol/l who would be found if various criteria were used to determine which people should be screened. The two most helpful criteria are a family history of CHD and obesity, each of which identifies roughly half of those with raised levels of cholesterol. Using all the criteria listed, it is possible to detect about 78 per cent of those with raised levels, but in order to do so it would be necessary to screen 66 per cent of the population. It is perhaps surprising that a family history of CHD, and, in particular, a family history of premature CHD, is not more helpful as a predictor of raised cholesterol levels in the general population, although it is a better predictor of familial hypercholesterolaemia, which, as mentioned before, carries a particularly high risk. One possible reason for a lower than expected pick-up related to family history is that when inheritance of the raised lipid levels occurs via the mother she may die from other causes before CHD manifests itself. This is because women with hyperlipidaemia have a lower risk of CHD compared with men and it tends to occur later in life.

Table 11.3. Detection of hyperlipidaemic individuals according to criteria used for screening

Screening criterion	Percentage of population with this attribute	Cholesterol > 6.5 mmol/l: Percentage found using this screening criterion	Cholesterol > 6.5 mmol/l: % of all people with cholesterol >6.5
Family history CHD	38	29	44*
Corneal arcus	6	40	10
Xanthomas	3	37	4
Obesity (BMI > 25)	43	32	54
Hypertension	12	38	19
Either family history or obesity	64	29	74
Any of the above	66	29	78

Data from the National Lipid Screening Project, Mann *et al.* (1983). *British Medical Journal*, **296**, 1702–6.

* If one screens everyone with a family history of CHD, one will find 29 per cent of people to have a cholesterol >6.5 mmol/l. This represents 44 per cent of the total population who have a cholesterol above this.

Table 11.4. Detection of hypercholesterolaemia levels > than 8.0 mmol/l) using various selective screening criteria

Screening criterion	Hypercholesterolaemia (>8.0 mmol/l) detected
Family history CHD	50%
Family history CHD under age 50	12%
Corneal arcus	13%
Xanthomas or xanthelasma	5%
Obesity (BMI>25)	54%
Hypertension (BP 160/90)	21%
Family history or obesity	77%
Any of the above criteria	81%
None of the above criteria	19%

A number of options are available to detect those with markedly raised cholesterol levels and these are outlined below.

General population screening

A quarter of the population with hyperlipidaemia would not fit into any of the categories suggested for special screening. Clearly the only way one could find all those at particular risk would be by general population screening. This has been recommended in some countries, but while this might well be the ultimate goal there are several major problems which need to be overcome:

1. At present, lipid measurements are usually made on a venous blood sample in a routine laboratory. This analysis takes time and the administration of such a screening service is a fairly complicated procedure. Equipment is now available for the measurement of cholesterol and triglyceride to be made on a finger-prick blood sample. Machines such as the 'Reflotron' (Fig. 11.3) are, however, costly and, although relatively robust, must be used carefully and

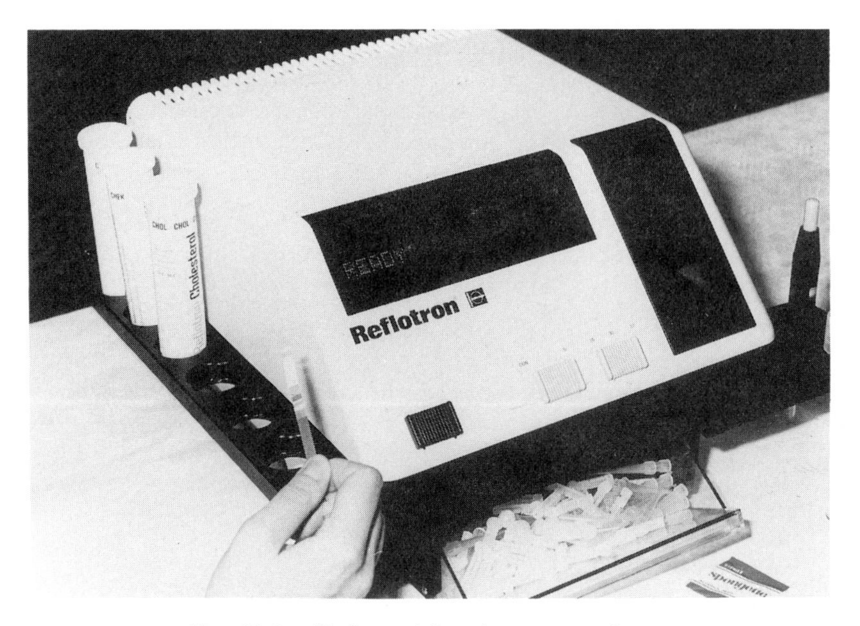

Fig. 11.3. 'Reflotron' dry chemistry analyser.

with consideration of quality control. At present few general practices are likely to invest in equipment costing around £3000, even though it can be used to make other biochemical estimations as well. Smaller and cheaper devices are being developed to overcome this problem. Cost–benefit calculations are rather difficult, but the availability of a quick and cheap method of measurement would improve the balance.

2. Although there is clear evidence for the benefit of cholesterol lowering, there is no data to show how many of those shown to have raised cholesterol levels in routine screening will accept the dietary advice (and sometimes drug treatment) necessary to reduce their blood lipids. Such data are being collected.
3. The response to the offer of general population screening seems to be relatively low. In Oxford, where awareness is high, about 60 per cent of those offered screening by letter of invitation have accepted, but this percentage is likely to be lower elsewhere.
4. There are insufficient numbers of doctors with adequate knowledge of disorders of lipid metabolism to carry out the further

investigation and treatment of those with familial disorders and other disorders not responding to dietary therapy. This situation is improving as a result of the recent evidence concerning the benefits of cholesterol lowering and the efforts of various organizations including the British Hyperlipidaemia and Family Heart Associations.

Much more progress is still needed on these four issues.

Case finding

It is now an accepted part of good clinical practice (although it is far from universally applied) that people with certain characteristics should be screened for hyperlipidaemia.

Screening those with a family history of CHD and those who are appreciably overweight will identify a reasonable proportion of those with raised lipid levels. The presence of premature corneal arcus or xanthomas are relatively specific indicators of hyperlipidaemia and should prompt blood lipid measurement, even though they will only identify a small proportion of hyperlipidaemic individuals. The presence of other risk factors for CHD (hypertension, diabetes) is also an indication for screening not only because such people may have elevated lipid levels, but also because the co-existence of more than one risk factor appreciably increases CHD risk. Anyone who is sufficiently interested to ask for his cholesterol to be measured should probably be given the opportunity since such people are likely to be well motivated to implement dietary changes or to comply with drug therapy should this be necessary.

Development to opportunistic screening

Machines similar to the Reflotron are likely to become more widely available and cheaper. Provided these are found to be accurate and reliable, and the facilities are developing for ensuring the necessary standardization procedures, we believe that they could enhance the extension of the approach to opportunistic screening. This involves offering screening to all who attend general practitioners' surgeries regardless of the purpose of the visit. Since 90 per cent of people in Britain attend their doctors' surgeries at least once over a 5-year period, this would enable a large proportion of the population to be screened. Although many young and middle-aged men do not attend in this period, they can often be contacted via their wives. Such a scheme has been successfully introduced into a number of general practices in Oxford. Here, where there is much enthusiasm, about 95 per cent of people have accepted the offer of screening when it is made directly at the surgery, despite the fact that it involves returning for the measurement of blood lipids and assessment of other cardiovascular risk factors.

In general it is usually appropriate to screen in the 25–59-year age group, although screening individuals for suspected familial hypercholesterolaemia may be undertaken in childhood. The way in which such a screening programme might be introduced into general practice is described in Chapter 12. Assuming that it is done as part of a cardiovascular risk assessment programme it is estimated that in an average British Health District about one eighth of the people in the appropriate age range are likely to be screened each year. Under these circumstances lipid screening might cost such a Health District about £100 000 per annum, (1988 costs), a sum equivalent to the cost of ten coronary artery bypass grafts.

Most patients identified should be handled in general practice as the majority will respond if they adhere to an appropriate lipid-lowering diet and for many others the drug therapy is relatively straightforward. A specialist clinic will, nevertheless, be needed to deal with difficult cases and to provide general assistance, and the aim should be for each District to have a lipid clinic. An analogy with diabetes may be helpful: there are more hyperlipidaemics than diabetics in affluent societies, and the evidence for benefit of treatment is probably greater, yet while most Districts have a specialist diabetes service, lipid clinics are few and far between. The high-risk strategy is likely to benefit the individuals concerned, but, in addition, it has considerable educational potential for the medical and associated professions as well as the public.

There is no doubt that the population and high-risk strategies must be implemented jointly if a serious attempt is to be made to reduce the CHD epidemic.

Evidence for benefit from an overall approach

In recent years mortality from CHD has fallen considerably in a number of countries most notably the U.S.A., Australia, and Finland (Fig. 11.4). It is impossible to establish the explanation for these changing trends with certainty, but they have occurred in parallel with a national will to

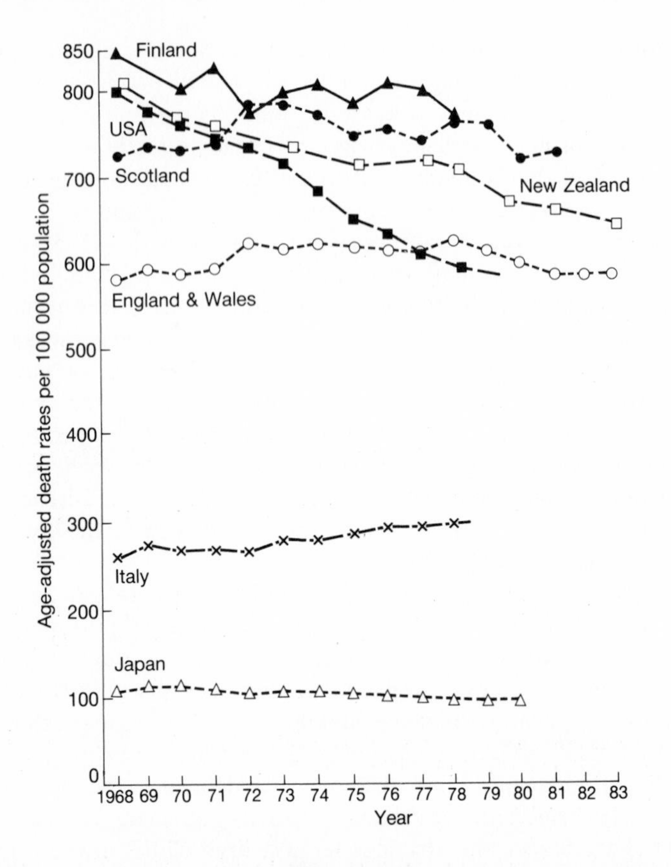

Fig. 11.4. Changing trends in CHD mortality in different countries (from Marmot 1984)

reduce CHD, dietary change designed to lower blood lipids, a reduction in cigarette smoking, and, perhaps, an increase in physical activity. In Finland, CHD has shown a greater fall in North Karelia, than in the rest of Finland (Fig. 11.5). The North Karelia province originally had among the highest CHD rates in Finland and a special CHD intervention project was introduced. It was originally intended to have an adjacent province as a control group, but it was impossible to prevent lifestyle changes in other provinces in a country so aware of its high CHD risk.

These trends do not provide absolute proof of benefit for a particular

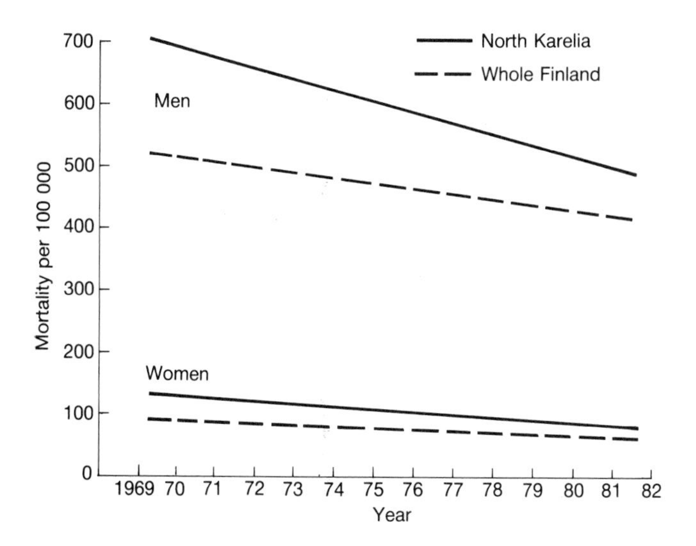

Fig. 11.5. Reduction in CHD in North Karelia compared with the rest of Finland. (Tuomilheto, J. *et al.* (1986). *British Medical Journal*, **293**, 1068–77.)

lifestyle change, but it is encouraging that the reduction of CHD has been particularly obvious in many countries and communities where special efforts have been made.

Mortality rates from CHD in males aged under 69 in different countries

Decreasing	**Increasing/Static**
USA	UK and N. Ireland
Australia	France
Finland	Denmark
Canada	Hungary
New Zealand	Bulgaria
Netherlands	Poland

Summary

Implementation of the population strategy for CHD prevention is likely to reduce the proportion of people in the population at moderate and high risk. Nevertheless, even with the widespread implementation of dietary advice aimed at reducing population cholesterol concentrations an appreciable proportion will remain at risk. This group will include those with familial hyperlipidaemias and others whose hyperlipidaemia responds inadequately to simple dietary advice. Such people will only be substantially helped by the high-risk approach. A high-risk approach should involve the screening of all those with a family history of CHD or hyperlipidaemia, appreciable obesity, clinical stigmata of hyperlipidaemia, or any other risk factors for CHD, and ideally doctors should be educated to look for lipid deposits as part of any physical examination.

Extension to offering opportunistic screening to everyone attending the general practitioner's surgery, regardless of the purpose of the visit, would greatly improve the identification of people with hyperlipidaemia.

12 Screening in general practice

In order to justify the case for screening, certain criteria must be considered. Probably the most important of these is that the disorder is of considerable severity and that effective treatment is available for those detected. Most people now consider that these criteria are met for CHD, and that the detection of risk factors is warranted. This would seem particularly true for the inherited conditions such as familial hypercholesterolaemia where the risks of CHD are so high (Fig. 12.1). In fact, a recent government white paper has suggested that general screening for risk factors should be implemented. The following chapter is just a suggested outline compiled by Dr I. Robertson and based on discussion and work with a group of general practitioners.

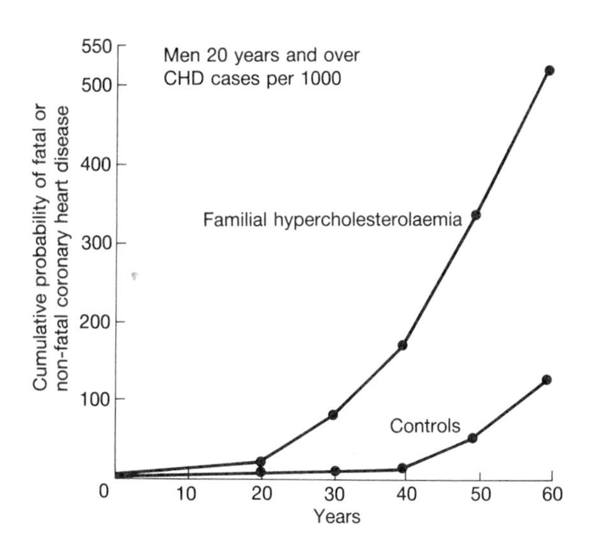

Fig. 12.1. Inherited lipid disorders and CHD.

Assessment of problem and overall plan

Decisions need to be made about:

(1) who one intends to screen—sex, age, conditions;
(2) how often one will screen;
(3) which factors one is going to screen for;
(4) how the process is going to be arranged—with regard to organization, patient contact, record keeping, etc.

Initial assessment:

• *Question about*—symptoms or history of CHD; smoking, alcohol, diet; family history of CHD, especially premature CHD.
• *Look for*—tendon xanthomas/xanthelasma/early corneal arcus.
• *Measure*—weight; blood-pressure; random plasma cholesterol; consider fasting cholesterol plus triglycerides if personal or family history of premature CHD, xanthomas present, or if patient has hypertension, diabetes or obesity.
• *Record*—flag notes of patients who are hypertensive, diabetic, hyperlipidaemic, and possibly smokers for follow-up.

For patients found to have a problem; Negotiation with the patient is needed to explain the problem, set objectives, and outline time limits in which to achieve these objectives. The initial risk-factor status and the plan of follow-up should then be recorded.

Follow-up is required to assess the degree of risk-factor reduction, to update advice and to ensure that any improvement is maintained. If possible a procedure should also be organized to recall defaulters.

Resource requirements

The initial assessment requires a relatively small amount of time and can be performed by a practice nurse. This can be arranged as planned sessions on a time-table, or offered opportunistically to patients attending for other reasons. It can be combined with checking cervical cytology status in women and tetanus immunization in everyone—activities which are likely to generate income for the practice.

Organization of screening on any scale does, however, require energy and time to plan and to monitor, and staff prepared to perform the day-to-day work.

Counselling and follow-up of patients with risk factors will be time consuming, and some training of staff may be required for this to have the optimum effect. Help with training may be available through local District Health Authorities.

Why screen for risk factors?

The most important reason is that patients want such a service. Opinion polls show that a regular physical examination is regarded by a large section of the public as a very important GP service. Many people wish to be told that they are fit and healthy, or if they have an early illness that can be cured. Medical opinion has generally held such general examinations not to be cost-effective, but for specific diseases such an approach is appropriate. This is particularly the case when the disease is severe, and effective treatment is available if it is detected early. Thus antenatal screening, and cervical and breast cancer screening are performed, for example.

CHD is a common, life-threatening, and disabling disease which can be difficult to treat, and potentially dangerous and expensive to try to cure. Drug treatment may reduce the symptomatology, but may not alleviate it, and coronary artery bypass grafting is often only partially effective. There are, however, some clearly identifiable factors which dramatically increase the incidence of CHD. Modification of these risk factors can reduce mortality from this disease (see Chapter 4). It is appropriate that this should be done before the patient develops symptoms. Identification of CHD risk factors requires no elaborate technology and treatment involves explanation, information, persuasion, and occasionally the judicious use of drugs—all the stock-in-trade of general practice.

General practice is the ideal location for a screening and advice programme. GPs have ready access to most of the patients on their list. Patients regard the issue as a legitimate one to be raised in general practice and the advice will be given credence if it comes from the practice, and particularly the GP. For these reasons more GPs need to perform opportunistic screening for CHD risk factors, just as some practices have paediatric and well-woman clinics to provide more systematic screening. These clinics are often conducted by practice nurses, with referral of problem cases to the GP.

How should the screening be organized?

Each practice has to decide what form any screening system will take after considering the special interests and skills of the individual doctors and nurses in the practice. A separate clinic for CHD risk-factor screening has certain advantages. Such activity is then kept separate from the diagnosis and treatment of acute illness, thereby preventing one set of instructions and advice interfering with another. Screening clinics can be run by nurses who are generally much better at gathering complete sets of data. They are also likely to be able to spend longer periods of time listening to the person, exploring their attitudes and thereby assisting them to come to the decision to change an unhealthy lifestyle. This leaves the GP with more time for problem cases and complex diagnostic problems.

A disadvantage of such a system, however, may be that the nurse lacks the doctor's authority and charisma, and therefore may be less effective at persuasion. It is thus important that the doctors are seen to be involved in any screening clinic, and are available to reinforce any message that is not getting across. The practice nurse will also need to possess or learn some counselling skills to increase effectiveness.

Who should be screened for CHD risk factors?

It is important to concentrate the effort of screening on those groups who are at greatest need and most likely to respond to advice. It is suggested that all those between 25 and 60 years of age should be screened (i.e. about a third of the patients on a general practitioner's list).

People younger than 25 are less likely to have risk factors such as hypertension, and they may not respond well to a suggestion of screening, or the type of advice likely to be given, since they are generally less concerned with their future health. In addition, some of the inherited hyperlipidaemias may not be manifest before this age.

Those over 60 years may be screened, but the emphasis in this group will be different—cholesterol measurement is not pursued as vigorously, while identification of hypertension and stroke prevention become more important. In the very elderly (over 75) it is likely that CHD prevention has become irrelevant, although hypertension may be worth treating because of the risk of cerebrovascular accident. In general, questions concerning social contacts, vision, hearing, and mobility problems may be more usefully explored.

Both men and women should be screened. Although women are not as commonly affected at a young age, they do suffer from CHD. They also cook most of the food, and so health promotion advice given to them will influence a wider group of people—both their husbands and their children. They will also help persuade their husbands to present for screening.

One of the priorities of health promotion is to counteract the health disadvantages of the manual workers and their families. Special efforts need to be made to persuade these groups to change their lifestyles, and this is difficult. These people generally have higher rates of risk and therefore more is required from them. They also have fewer resources so that increasing exercise or changing their diet will represent an increase in actual or perceived costs which they may feel unable to afford. For some people the day-to-day struggle may be so great that they will not accept the need to make changes now for long-term benefits. Some will have beliefs about their health which will be totally contrary to the advice being given. These beliefs and motivations must be explored and reasons for lifestyle changes found in the group before a major reduction in CHD mortality can be achieved.

How should patients be contacted?

All patients screened should ideally be volunteers. Thus a letter of invitation to the screening clinic can be given by the receptionist to a patient when they attend the surgery for other matters. Up to 70 per cent of patients attend their GP each year; and some 80 per cent in three years and this approach allows an orderly introduction of a new service preventing any initial rush, and saves the cost of widespread postal invitations.

When an individual is found to have a particularly raised cholesterol, then an effort should be made to determine whether this is a familial disorder, such as familial hypercholesterolaemia. The stigmata which indicate this condition are given in Figure 12.2. The reason for specifically trying to identify these particular individuals is that their risk is very high (Table 12.1), and because the diagnosis and subsequent

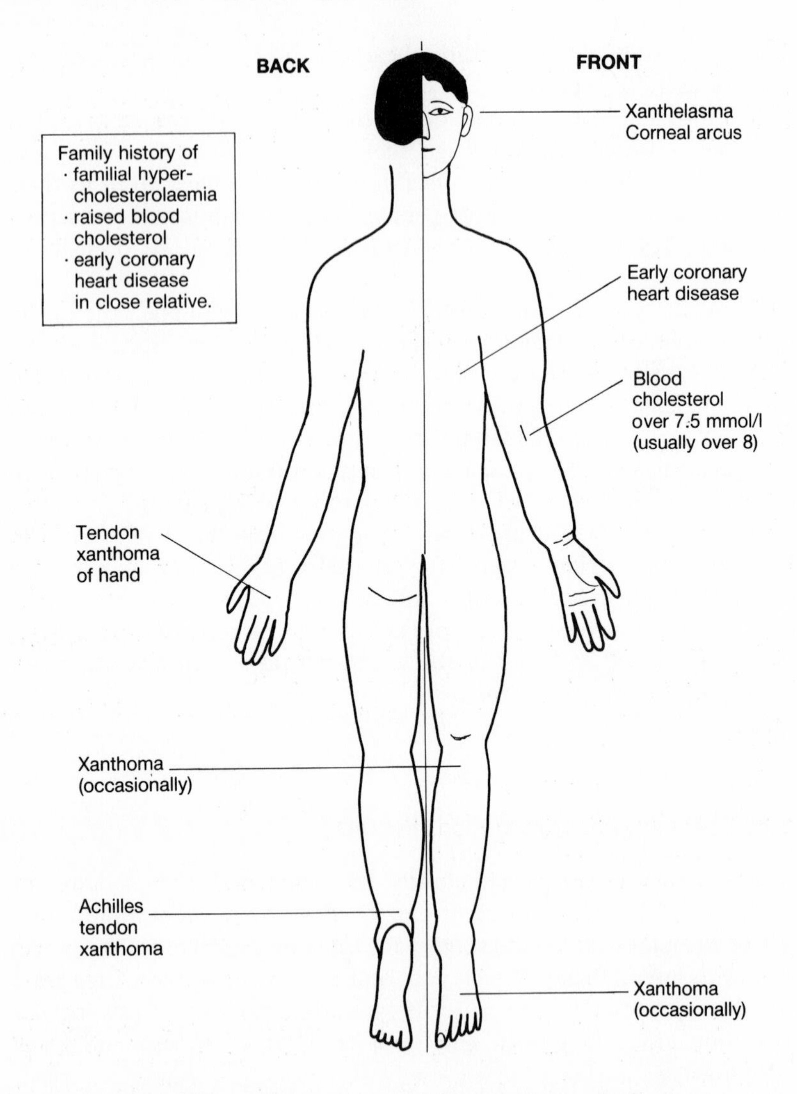

Fig. 12.2. Features of familial hypercholesterolaemia.

prompt screening of all their first-degree relatives may identify a large number of affected individuals. As more and more people are screened, the message that the clinic exists is passed by word of mouth. The number of unscreened individuals attending the surgery thus diminishes

Table 12.1. Percentage of cases of heterozygous familial hypercholesterolaemia showing symptoms of CHD, myocardial infarction (MI), or dying of CHD by age and sex, (From Goldstein and Brown, 1983; Slack 1969) giving an idea of the severity of the problem. (Note—Many untreated.)

	Males			Females		
Age to (years)	CHD symptoms	MI	CHD death	CHD symptoms	MI	CHD death
30		5%	—	3%	—	—
40	20%		—			
50	45%	51%	25%	20%	12%	2%
60	75%	85%	50%	45%	51%	15%

and it may be necessary to send out letters of invitation. This will allow the targeting, for example, of working class men.

A rough estimate of the potential findings in a practice of 10 000 are as follows:

- familial hypercholesterolaemia 20
- other familial hyperlipidaemias 25
- plasma cholesterol >7.5 mmol/l 200
- hypertension (patients under 65) 200

Record keeping

The record card forms a useful health summary which, along with a treatment summary and repeat prescription card, can form an extremely useful reminder at the front of the notes of the overall status of the patient. It is also useful to flag the front of the notes as a prompt to the receptionist and anyone else, to offer the patient a screening appointment when next they attend surgery. A record of the date of screening should be kept in order to invite the patient back for re-screening for some of the risk factors five years later.

Table 12.2. Possible action plan for screening

Risk factor	Action	Follow-up
Smoker	— Counsel on adverse effects Give advice on ceasing.	Follow-up if possible
Family history of CHD	— Check fasting lipids	
Obese	— Check BP, plasma lipids. Advise diet, arrange dietitian visit if appropriate. Suggest joining slimming club.	Follow-up (nurse/dietitian)
No exercise	— Suggest increase in regular exercise if no contraindications. Encourage walking, swimming etc.	
Hypertension	— BP diastolic 90 or less.	Check in 5 years.
	— Diastolic 90–114 mmHg.	Check a few weeks later.
	— Diastolic 115+ or systolic >160	Refer to doctor
	Check U & E, urine protein, retina. ECG Diet if obese. Stress reduction Arrange for further visit. Consider drug treatment if remains elevated.	Regular follow-up

Diabetic	— Check BP, lipids, renal function, proteinuria, eyes, feet, method of glucose monitoring.	Regular follow-up
Plasma lipids	— If raised consider: Weight reduction Exclusion of secondary hyperlipiaemia Diet Refer to doctor and consideration of drug treatment if levels remain high on appropriate diet. Screen family if index patient appears to have a familial hyperlipidaemia. Consider referral to lipid clinic if any problems.	Follow-up of response

Monitoring the effectiveness of screening

It is not sufficient just to identify CHD risk factors and give advice on lifestyle changes. Staff need to know that as a result the patients do give up smoking, or take exercise and change their diet. Some patients will not understand the message, or not get around to doing anything about it. Such failures may teach the practice staff how to modify the message for different types of people.

Success in this field, and the ability to demonstrate this success, is extremely important to the health of the nation. For general practice to be able to demonstrate its prime role in this success would strongly enhance its prestige and enhance its role in the running of the health services of this country. The effort is great, but so are the rewards.

Section IV

13 Practical help

The FHA

Several years ago, a number of patients with familial hypercholesterolaemia (FH) became very aware of the lack of information generally available. These individuals, assisted by some of their consultants, approached the Simon Broome Heart Research Trust. A self-help group —the Familial Hyperlipidaemia Association (FHA) now the Family Heart Association was thus formed. The aims of the Association are:

1. To make the public and the medical profession more aware of this important and fairly common condition.
2. To inform and support those found to have FH.
3. To encourage further research into the cause and treatment of the condition.

Their initiative encouraged the Simon Broome Trust to produce an informative booklet for patients, with the help of a number of experts on lipid disorders. An excerpt is given later. It is available on request from:

The Hon. Secretary, FHA, Box 116, Kidlington, Oxfordshire, OX5 1DT. (Telephone no. 0865 79125).

Since then the FHA have produced several other booklets for patients and for general practitioners. The Association encourages all people with FH and other inherited lipid disorders to join, and receive and contribute suggestions, recipes etc., which are circulated in a newsletter. Anyone interested can join as an associate member, and receive this literature.

Much of the work performed with new members is answering queries and counselling, and a network of regional counsellors has now been established. Considerable time may need to be spent supporting new patients, helping them to come to terms with the problem and to adopt a positive approach. The FHA can also act as a resource for general practitioners as they have built up contacts with interested GPs, lipid clinic staff, and groups involved with CHD prevention and care.

The following is an example of one of the booklets available from the

FHA for people with familial hypercholesterolaemia. Simpler booklets and dietary advice leaflets are also available.

Understanding FH

What is familial hypercholesterolaemia?

Hypercholesterolaemia is the medical name given to the condition of high blood cholesterol. 'Hyper' means raised; Cholesterol is the fatty substance described below, and 'aemia' means in the blood. People with high blood cholesterol tend to have a greater risk of developing coronary heart disease.

Familial hypercholesterolaemia (FH)

Is the name given to high blood cholesterol where it is inherited and passed from parent to child.

What is cholesterol?

Cholesterol is one of the fatty substances present in the blood and all body tissues. It is an important part of the outer envelopes of each cell in the body. Cholesterol comes partly from the diet, and is also made by the body.

What is high blood cholesterol?

People with high blood cholesterol have an amount of cholesterol in their blood which is higher than normal for people in this country.

The body normally maintains a delicate balance between the cholesterol eaten, the cholesterol made in the liver, and the amount required by the body so that supply matches demand. In FH there is an inherited abnormality in the cells of the body which upsets this balance so that the level of cholesterol in the blood is high and there is a gradual build-up of fatty cholesterol deposits in the body.

People with high blood cholesterol tend to have a greater risk of developing coronary heart disease.

What is coronary heart disease?

Coronary heart disease is the name given to the narrowing of the coronary arteries which convey blood to the heart. The narrowing is caused by a build-up of fatty cholesterol deposits which, in time, could severely restrict the amount of blood reaching the heart. If this happens suddenly there may be damage to part of the heart muscle—this is known as a heart attack (coronary thrombosis). With FH this can happen in relatively young people.

Statistics show that men run a greater risk of developing coronary heart disease than women. There are also other factors apart from high blood cholesterol that increase the risk of developing heart disease. These include smoking, lack of exercise, over-eating leading to obesity, diabetes, and high blood-pressure.

Diagnosing FH

How can you tell if you have FH?

If a member of your family, whether one of your parents, uncles, aunts, brothers, or sisters, has had a heart attack (coronary thrombosis) at a young age—in early or middle adult life—this could be because of FH. Other indications of the condition are:

1. Xanthomas—these are swellings in the tendons on the back of the hands or ankles as a result of a build-up of cholesterol here.
2. Corneal arcus—a white band towards the end of the coloured part of the eye in younger people (under 50 years). The band may occur without FH, and indeed without any change in cholesterol, in older people.
3. Xanthelasmas—yellow lumps or streaks of fat in the skin close to the eye.
4. High blood cholesterol in a close relative.

If any of these apply to you, you should first consult your family doctor and have your blood cholesterol and other fats measured.

The outward signs of FH are not present in all patients, and even when they are they may cause little or no discomfort or inconvenience. However, they do help the doctor in making the diagnosis.

Not everyone with high blood cholesterol has FH, but it makes a good deal of sense to have the tests.

If someone in your family is known to have FH then **IT IS VERY IMPORTANT** that all members of the family ask their doctor to check their blood cholesterol. This test is available on the NHS.

Can FH miss a generation then reappear in grandchildren?

FH does not miss a generation, but because it does not always result in a heart attack at an early age, it may sometimes not be diagnosed if the blood cholesterol is not measured.

Reducing the risk

Can the risk of heart attacks be removed or reduced if blood cholesterol can be lowered?

High blood cholesterol increases the risk. Lowering blood cholesterol is the essential basis for treatment. Once FH is diagnosed, treatment should be started as soon as possible, before disease has developed in the coronary arteries.

Diet is a very important factor in lowering cholesterol levels in the blood.

Must I always keep to my diet?

Departures from the diet on special occasions will not cause problems, provided you normally keep carefully to your diet. It is the average diet over a period of time that is important. Remember, though, that there is a danger of finding too many excuses for special occasions!

How can I make my children's diet more interesting?

It is difficult for children to keep to a special diet which excludes chocolate, cakes, and crisps among other things. The following tips may help.

Sandwiches can be made with tuna, salmon and cucumber, chicken, lean roast beef and ham, cottage cheese, and other low fat cheeses, with tomato, pickle, pineapple, yeast extract, jam, and honey. You can also pack individual salads in a plastic container. Add a piece of fruit, carton of yoghurt, muesli biscuits, slice of bran and banana loaf, some dried fruit.

Encourage your children to eat more vegetables. Try cooking the vegetables in different ways. Make the children 'polyunsaturated' crisps. Using a potato peeler, cut slices very finely, rinse under the tap, pat dry, and deep fry in polyunsaturated oil. These are much cheaper than commercially bought ones.

What if diet alone is not enough to lower my blood cholesterol to normal?

In many people with FH, diet, however strict, will not lower the blood cholesterol sufficiently and drugs have to be used.

There are two main types of drugs:

1. A powder (mixed with water or fruit juice) which you swallow. It is not absorbed by the body and works by preventing the absorption of a fluid called bile (which is made from cholesterol). The body then makes more bile and so uses up body cholesterol and the level in the blood falls. Examples of such drugs are cholestyramine (Questran) and colestipol (Colestid).
2. A capsule or tablet which is absorbed by the body, where it acts to cut down the manufacture of cholesterol by the liver. Examples of this type of drug are clofibrate (Atromid-5), bezafibrate (Bezalip), probucol (Lurselle), and various nicotinic acid preparations.

Sometimes a combination of drugs is more helpful than one alone, and all drugs work better if the diet is continued.

Are there any known side-effects of the drugs?

All drugs can occasionally cause side-effects. However, in practice the drugs used to treat FH rarely cause serious problems. Any side-effects should be discussed in detail with your doctor in relation to the particular drugs you are taking.

Some of the drugs (especially the first type) may alter bowel habits, but this is nearly always a temporary effect. If you are on a high-fibre diet, bowel upset (constipation) is less likely to occur.

What about my weight?

People with FH should try to avoid being overweight because it is more difficult to keep blood cholesterol down. Also being overweight itself increases the risk of developing heart disease.

What about alcohol?

Yes, it is fine to have a social drink, but remember, if you have a weight problem, that it is fattening.

Should I continue to smoke?

No, smoking is a disaster! Smoking is already associated with a high risk of heart attacks so it is particularly dangerous when FH has been diagnosed.

Is it safe to take exercise—or ever play sports?

An active lifestyle, including pursuits such as walking, cycling, swimming, and gardening, is to be encouraged in helping to prevent coronary heart disease. However, if you have not previously been active you should take-up exercise very gradually. In adults with FH who may have some narrowing of the coronary arteries without symptoms of heart disease, sudden vigorous exertion and high competitive physical sports are probably best avoided. Discuss your exercise plans with your doctor.

Information for doctors

Other booklets on CHD risk factors are available from the British Heart Foundation, 102 Gloucester Place, London W1. Pamphlets on FH are available from the FHA. Up to date readable reviews on lipid topics can be found in Lipid Reviews, which is available free and published by Current Medical Literature Ltd., 40 Osnaburgh Street, London NW1.

Public awareness

Community attitudes to people who wish to lead a different lifestyle from the generally accepted norm can be hostile. This can apply to people who wish to change their life-style because of hyperlipidaemia. Individuals may find resentment to their requests for 'low-fat' foods in work canteens, and parents often complain that they have difficulty over school meals for their children.

Health professionals and the public need more education about the importance of hyperlipidaemia and other CHD risk factors, and their treatment. This way a gradual change in attitudes can hopefully be achieved. The food industry in general, both manufacturers and retailers, have shown little concern about the need for a prudent diet which would reduce the risk of CHD for the whole population, as well as special high-risk groups. The food chain is complex and includes producers, processors, manufacturers, distributers, caterers, government departments, and the consumer. Complicated and sometimes conflicting interests are involved. Changes are really needed in the agricultural policies of this country, for example, to encourage farmers to produce leaner meat, to reduce dairy farming, and to increase the production of cereals, vegetables, and fruit. This would of course have far reaching implications.

The situation is improving slowly. Information on diet and CHD is now quite widely available, and some school courses are including more information on nutrition. Much still remains to be done, however, particularly within certain social groups. Some food manufacturers are realizing that there is a growing market for 'health-promoting' foods. Several supermarket chains now stock low-fat, high-fibre foods which are well labelled, and provide leaflets and information on healthy eating. Alternatives of packaged or tinned food are available without added sugar, salt, and preservatives, in some shops. The information on processed foods is generally improving, but there is still a long way to go. Food labelling needs to be dramatically improved to include the fat content, its source and the percentage of saturated and unsaturated fat, as well as other nutritional information.

RECIPES

The following section includes a selection of recipes for low-fat meals.

MAIN COURSES

1 Chicken, ham, and apple salad

Serves 4,
Each serving: 210 kcal, 10 g fibre, 4 g fat

2 medium sized apples (red or green, cored, and diced)
100 g (3½ oz) cooked chicken or turkey, diced
100 g (3½ oz) cooked lean ham, diced
200 g (7 oz) cooked peas
3 large stalks of celery, chopped
30 g (1 oz) sultanas
Lettuce

Mix ingredients except lettuce—lay on bed of lettuce. Use 50 ml (¼ pint) of low fat plain yoghurt plus seasoning as dressing.
Serve with cooked brown rice (½ cup per person) mixed with sweet-corn kernels.

2 Country vegetable risotto

Serves 4
Each serving: 440 kcal, 23 g fibre, 1 g fat

200 g (7 oz) haricot beans or kidney beans
1 tbsp polyunsaturated oil
1 medium-sized onion, chopped
1 clove garlic (optional)
3 large carrots, diced
⅓ tsp salt and pepper

1 tbsp marjoram or sage
400 ml (14 fl oz) stock
200 g (7 oz) long-grain brown rice
200 g (7 oz) fresh or frozen peas
3 tbsp chopped parsley
1 bunch spring onions, green and white parts, sliced
30 g (1 oz) parmesan cheese

Cook the beans or use drained tinned beans. Heat the oil in a saucepan and fry the onion and garlic for a few minutes. Add the carrots and swede and continue to cook, stirring occasionally for 3–4 minutes. Sprinkle the seasoning and marjoram or sage into the pan, cover with water and simmer for 20–25 minutes, or until almost tender. Drain.

Meanwhile, in another saucepan, bring the stock to the boil, sprinkle in the rice, stir, and return to the boil. Lower the heat, cover the pan, and cook gently for 30–35 minutes, adding more stock until the rice is almost tender and the stock absorbed. Add the beans, peas, and the cooked vegetables, and continue cooking for 10 minutes. Remove from the heat, adjust the seasoning, mix in half the parsley, pile onto a hot serving dish and garnish with the remaining parsley, spring onions, and cheese. The consistency should be moist and creamy.

An alternative method is to boil the beans until half-cooked, then add the vegetables and rice. As different kinds of beans and brown rice vary in the cooking time required, the first method ensures the beans are properly cooked. The alternative cooking method, however, produces a dish with a very low fat content.

Many other vegetables may be used—unpeeled diced potatoes, courgettes, broad beans, green beans, leeks etc.

3 White fish in white sauce

White fish (cod, haddock, plaice, coley, whiting, halibut)
2–4 oz per person
Skimmed milk
Potato powder
Parsley or pepper

Thaw fish and place in saucepan, cover with skimmed milk, about ⅓ pint per person and bring to the boil. Simmer for 30 seconds (for plaice) to 2 minutes (cod). Drain milk into basin and add potato powder until sauce is desired thickness, stirring continuously. Cover fish with sauce and sprinkle with parsley. Serve with boiled potatoes and vegetables of choice.

4 Pasta and meat/vegetable sauce and grated cheese

Pasta 50 g (2 oz) per person
Spaghetti, pasta whirls, or macaroni (preferably high fibre or wholemeal)
Small amount of diced ham, low fat mince, diced chicken or turkey.
Onion, peppers, mushrooms, courgettes, kidney beans, chick peas, or other similar pulses
Tin of tomatoes

Cook pasta according to packet instructions.
Fry vegetables in a small amount of polyunsaturated margarine.
Add tomatoes and simmer for 5 minutes.
Drain pasta, pour sauce over pasta. Sprinkle with grated or parmesan cheese and black pepper.

5 Chicken lasagne

Serves 4
Each serving; 380 kcal, 40 g (4 units) carbohydrate, 9 g fibre, 26 g protein, 13 g fat.

140 g (5 oz) wholemeal lasagne
15 ml (1 tbsp) corn oil
1 large onion, chopped
1 clove garlic, crushed
400 g (14 oz) tomatoes

1½ tbsp marjoram
4 outer stalks celery, diced and lightly cooked
6 tbsp chopped green pepper
170 g (6 oz) chopped chicken, diced
seasoning
30 g (1 oz) cheese, grated

Sauce:

1 small onion, chopped
carrot, chopped
turnip, chopped
425 ml (¾ pint) skimmed milk
30 g (3 tbsp) wholemeal flour
15 g (½ oz) polyunsaturated margarine
seasoning

Garnish: paprika

Heat the oven to 190°C/375°F/gas 5.
Cook the lasagne in boiling salted water until half-cooked, drain carefully. Heat the oil and cook the onion and garlic for 5 minutes, until soft. Add the tomatoes, marjoram, celery, peppers, chicken, and seasoning, and cook for 5 minutes. To make the sauce, simmer the vegetables, bay leaf and mace in nearly all the milk for 15 minutes. Remove the bay leaf and mace and mix the ingredients in a blender. Return to the pan and bring to the boil.

Meanwhile, blend the flour with the remaining milk. Stir the hot milk into the flour, add the margarine and seasonings, return to pan and cook gently for a few minutes. Place the chicken mixture, sauce, and lasagne in alternate layers in a dish, finishing with sauce. Scatter the cheese evenly over the top* and cook for 30–35 minutes. Sprinkle a little paprika on top and serve.

* This dish can be prepared in advance to this stage, cooked, covered and stored in the refrigerator, but allow 40–45 minutes for the final cooking.

6 Chilli con carné

Serves 4
Each serving: 530 kcal, 60 g (6 units) carbohydrates, 27 g fibre, 37 g protein, 16 g fat.

30 ml (2 tbsp) corn oil
2 medium-sized onions, chopped
1 clove garlic, crushed
170 g (6 oz) lean minced meat
20 g (2 tbsp) wholemeal flour
2 tbsp tomato purée
250 ml (9 fl oz) stock, hot
1–3 tsp chilli powder
400 g (14 oz) canned tomatoes
450 g (1 lb) red kidney beans, soaked or canned
salt
1 medium/large green pepper, chopped

Heat the oil and gently fry onions and garlic for 5 minutes. Toss the meat in flour, add to the onions and cook until brown, stirring. Mix tomato puree with the hot stock and gradually stir into the meat. Add the chilli powder, tomatoes, and beans, and boil for 10 minutes. Stir, cover tightly, and simmer gently for 1–1¼ hours until the beans are cooked. 10 minutes before serving stir in salt and the green pepper.

Use a mixture of meat and textured vegetable protein, or soya soaked in a small amount of hot water and marmite, in place of all meat. Instead of red kidney beans, use haricot, butter, or soya beans. The cooking time should be adjusted.

7 Vegetarian bean paella

Serves 4
Each serving: 420 kcal, 70 g (7 units) carbohydrate, 20 g fibre, 16 g protein, 10 g fat.

100 g (3½ oz) red kidney beans
60 g (2 oz) mung or haricot beans

225 g (8 oz) long-grain brown rice
450 ml (¾ pint) water
½ tsp turmeric (optional)
1 small aubergine, diced
30 ml (2 tbsp) corn oil
1 onion, chopped
1 stick celery chopped
1 medium large green pepper, chopped
2 large carrots, diced
300 g (10½ oz) canned tomatoes, drained
115 g (4 oz) button mushrooms
seasoning
2 tbsp chopped parsley
Saffron

Cook the beans. Meanwhile, put the rice, water, saffron and 1 tsp salt in a saucepan and cook for 30 minutes. Sprinkle salt over the aubergine, leave for 20 minutes, then wipe dry. Heat the oil in another saucepan and gently cook the onion and garlic for 5 minutes. Add the aubergine, celery, green pepper, and carrots and cook gently, stirring occasionally, for 10 minutes. Stir in the tomatoes and mushrooms, and continue cooking for 5 minutes. Gently mix the vegetables and beans into the rice, adjust the seasoning if necessary. Cover the pan and continue cooking gently for 15 minutes. Turn off the heat and keep the mixture warm for 10 minutes. Gently fork in the parsley and serve.

ACCOMPANIMENTS

8 Parsley rice

Serves 4
Each serving: 270 kcal, 6 g fibre, 2 g fat.

225 g (8 oz) long-grain brown rice
30 g (1 oz) polyunsaturated margarine

1 large bunch spring onions, sliced
4 stalks of celery
6 tbspn chopped parsley or other fresh herbs
seasoning

Cook the rice. Meanwhile, melt the margarine in a saucepan. Add the spring onions and celery and fry gently, stirring occasionally, for 7–10 minutes or until just tender. Add the freshly cooked, moist rice and chopped parsley, adjust the seasoning and allow to heat through thoroughly, stirring occasionally. Serve as an accompaniment to a main course.

9 Mackerel pâté

1 tin mackerel 150 g
1 tin butter beans 8 oz
Pepper

Mix mackerel, pepper, and butter beans together. Mash with fork or in a food processor (if available). Place in serving dish.

SWEETS

10 Apple oatmeal crumble

Serves 4–6
Each serving: 230 cal, 40 g (4 units) carbohydrate, 7 g fibre, 7 g fat.

570 g (1¼ lb) cooking apples, cored
¼–½ tsp ground cloves
low calorie or sugar-free sweetener to taste
60 ml (4 tbsp) hot water
30 g (4 tbsp) rolled oats
100 g (3½ oz) wholemeal flour
30 g (1 oz) polyunsaturated margarine

Heat the oven to 180°C/350°F/gas 4. Slice the applies in cross cut slices to avoid long strips of peel. Place in a baking dish, sprinkle with the cloves, and add the sweetener to taste. Mix together the oats and flour, and rub in the margarine. Sprinkle over the fruit and bake for 30–40 minutes.
Other fruits, such as rhubarb or plums, can be used.

11 Rhubarb charlotte

Serves 4
Each serving: 130 cal, 20 g (2 units) carbohydrates, 7 g fibre, 4 g fat.

450 g (1 lb) young rhubarb*
170 g (6 oz) wholemeal breadcrumbs
grated rind of 1 orange
½ tsp ground ginger
60 ml (4 tbsp) orange juice
¼ tsp ground cinnamon and nutmeg (optional)
low calorie or sugar-free liquid sweetener to taste
30 g (1 oz) low-fat spread

Heat the oven to 190°C/375°F/gas 5
Cut rhubarb in 2.5 cm (1 inch) lengths and place in a layer in a non-stick pie dish or casserole. Sprinkle a layer of breadcrumbs on top. Mix together orange rind, ginger cinnamon, and nutmeg, if on top. Mix together orange rind, ginger cinnamon, and nutmeg, if used, and sprinkle a little over the breadcrumbs. Repeat the layers, finishing with a layer of breadcrumbs, but before the final layer, pour over the orange juice mixed with sweetener. Dot the final layer of breadcrumbs with the low-fat spread. Cover and bake for 10–15 minutes until the top is crisp and lightly brown. Serve hot.

*Older rhubarb may be used, but stew gently with a little water to part-cook before assembling the pudding.

Some of these recipes are adapted from ones given in 'The Healthy Heart Diet Book' by Longstaff and Mann and the 'Diabetics' Diet Book' by J. Mann, published by Martin Dunitz.

Recipes Using Questran (cholestyramine)

Soft drinks

1 sachet of Questran
250 ml (8 fl oz) soft drink any flavour

Put Questran powder in a large glass. Add 3–4 oz soft drink and stir until all Questran is in suspension. This will cause the soft drink to foam. Wait until the foam has settled and add rest of soft drink, slowly. Occasionally stir the soft drink, gently. Low-calorie soft drinks should usually be used in place of sugar-sweetened drinks for adults.

Fruit juices

1 sachet of Questran
250 ml (8 fl oz) orange or grapefruit juice

Mix Questran with fruit juice stirring all the time. Add ice if desired. Variation: instead of using all fruit juice, replace 3 oz with soda water or lemonade.

Milk shakes

250 ml (8 fl oz) skimmed milk
2 teaspoons milk-shake powder
1 sachet of Questran

Combine the powders together in a glass. Gradually add the milk, stirring all the time.

Yoghurt drink

150 g (5 oz) raspberry yoghurt or any low-fat fruit flavoured yoghurt
125 ml (¼ pint) skimmed milk
1 sachet of Questran

Mix everything together and stir thoroughly until evenly dispersed.

Banana milk

1 large banana
250 ml (8 fl oz) skimmed milk
1 sachet Questran

Liquidize together. Serve.

Hot lemon

Use lemon juice or lemon squash or other fruit cordial.
First add hot water to make one cup, then add one Questran sachet and stir thoroughly until Questran is in suspension.

Ginger special

10 ml (4 fl oz) orange juice or lime juice
250 ml (8 fl oz) ginger ale or low-calorie ginger ale
2 sachets of Questran
Ice cubes if required

Sprinkle Questran into orange juice and stir thoroughly until Questran is evenly dispersed. Add ginger ale. Stir.

Porridge

Porridge oats (quantity as recommended by manufacturer)
Skimmed milk
1 sachet Questran

Prepare porridge according to package directions using skimmed milk and ¼ cup more water than directions suggest. When cooked remove from heat and sprinkle one sachet of Questran into an individual portion and mix thoroughly. If porridge is too thick, stir in more hot water.

Caribbean yoghurt breakfast

2×5 oz (150 g) cartons of low-fat natural or fruit yoghurt
1–2 tsp muscovado sugar (if required)

1 pinch cinnamon
1 tbsp sultanas
1 grapefruit (segmented)
1 tbsp of muesli
1 sachet Questran

Mix all ingredients except Questran. Divide into two servings and add Questran to individual portion, stir thoroughly. This recipe can be used with many other fruits—fresh orange, apple, melon, banana, or pineapple.

Chicken/turkey casserole

Serves 4

10 oz uncooked chicken (without skin)
14 oz mixed vegetables, carrots/sweetcorn/beans/peas/celery
1¾ pints boiling chicken stock
Black pepper/soy sauce seasoning

Mix ingredients and cook for 1½ hours

Dissolve Questran in small amount of warm water and mix in thoroughly with portion.

Some of these recipes are adapted from ones given in a booklet produced by Bristol Myers.

Eating Out, Packed Snacks, and Buffet Meals

Eating Out

If possible choose somewhere to eat where you know the menu will include simply cooked food rather than elaborate made-up dishes with unknown ingredients. Contact a number of restaurants in advance to find whether they have suitable meals.

Starters
Fruit, fruit juice, tomato juice.
Avoid pâté, cream soup, shellfish cocktail.

Main course
Fish; grilled or poached.
Poultry, roast or grilled meat, cold meat: trim off any obvious fat or skin.
Avoid sauces, stuffings, dumplings, pastry.

Vegetables
All salads, add dressing of oil and vinegar at the table.
Any vegetable prepared without fat or sauce.
Jacket or boiled potatoes.
Avoid salads already mixed together with rich creamy dressings, and creamed, roast, or fried vegetables.

Bread
Wholemeal bread or rolls, crispbreads.
Avoid butter.

Dessert
Fresh or canned fruit, jelly, sorbet.
Avoid cream, rich creamy ice creams, pies, and gateaux.

To drink
Fruit juice, mineral water, or moderate amounts of wine, beer, and spirits (if permitted). Coffee, tea.
Avoid cream in coffee or tea.

Remember If it is impossible to avoid some unsuitable food, have a small helping and satisfy your appetite with generous helpings of other foods. The occasional *small* lapse will not cause harm, but it is important to return immediately to your recommended dietary routine.

Packed Snacks

It is probably sensible to try and find suitable containers for certain foods.

Soups

Main course or light soups (low calorie or home made with low fat content) carried in a thermos flask on cold days.

Wholemeal rolls or sandwiches (in plastic snap bags) or crispbreads.

Spread sparingly with polyunsaturated margarine, low-fat margarine spreads or low-calorie salad dressings. Alternatively, just use a filling made very moist and tasty.

Fillings can include: slices of poultry, lean meat or ham—or mince these and mix with chutney, pickles or low-fat sauce; canned fish mixed with vinegar, low-fat salad dressing, tomato puree, or bottled sauce; cottage or low-fat curd cheese mixed with herbs, sauce, pickles, chopped salad vegetables, or chopped grapes; home-made pâté spread directly on to bread.

To the above fillings, add sliced or shredded salad ingredients and homemade salad dressings, or have carrots/celery/tomatoes separately.

In cartons

Salads of vegetables, brown rice, and pasta mixed with chopped poultry, lean ham, or fish, and permitted salad dressings.

Dessert

Fresh or dried fruit and nuts.
Cold desserts in seal-top containers. Low-fat yoghurt.
Biscuits baked with appropriate ingredients.

To drink

Cold drinks in seal-top tumblers. Coffee or tea with skimmed milk.
Sometimes: canned beer, bottle or cartons of wine, or non-alcoholic wine.

Catering for Buffet Meals

Select from the following foods:

Starters
Melon, grapefruit, fruit juice.
Home-made pâté (see recipes) with wholemeal melba toast.
Mixed vegetable broth with added wholemeal cereals, pulse soups, and croutons of wholemeal bread, toasted or fried in polyunsaturated oil.

Main courses
Moderate amounts of fish, meat, and cheese extended by combining with wholemeal cereals and vegetables, and presented to make a variety of dishes. For example, flans, pizzas, risottos, pasta with succulent meat or fish sauces, casseroles with pulses and other vegetables, salmon (or other fish) mousse.

Vegetables
Colourful salads made from summer and winter vegetables combined with pulses, brown rice, and pastas.
Crisply cooked vegetables.
Rice/sweetcorn/pepper mixes.
Potatoes baked or boiled in skins with low-fat cheese or dressing.

Dressings
Home-made French or yoghurt dressings.
Proprietary low-calorie dressings.

Bread
Wholemeal wheat bread, rye bread, and rolls, crispbreads, oatcakes.
French bread sticks warmed in the oven.

Spreads
Tubs of polyunsaturated margarine, low-fat margarine spread and home-made, low-calorie spreads. Some polyunsaturated margarines and spreads may be made into 'pats' using a butter curler.

Desserts
Fresh fruit.
Fresh fruit made up into fruit jellies and flans.
Fruity puddings such as charlottes and pies made with appropriate thin pastry (polyunsaturated fat and half brown/half white flour).
Sorbets, home-made ice cream.

Information sheet for patients on dietary principles

What part does diet play in the treatment of hyperlipidaemia (high blood fats)?

Dietary modification is the first line of treatment in hyperlipidaemia. The ideal diet is one which aims to achieve ideal body weight, is low in fat, and has a relatively high ratio of polyunsaturated fat to saturated fat, and a high content of dietary fibre. In the typical affluent Western diet, approximately 40 per cent of all calories come from fat, 45 per cent from carbohydrate, and 15 per cent from protein. For those with hyperlipidaemia, fat should be reduced to less than 30 per cent of total calories. This should be done by decreasing saturated fat. Fibre-rich carbohydrate should be increased. This way of eating has been endorsed by official recommendations throughout the world as a major way of reducing the epidemic of coronary heart disease. It is suitable for the whole family whether or not they have hyperlipidaemia. Those who do not have raised lipid levels can be less strict.

Important points about your diet

These general guidelines will help you to follow a healthy balanced eating pattern. The same dietary principles apply to all types of hyperlipidaemia, though some aspects may be particularly emphasized for people with certain conditions.

Eat less fat. It is important to eat less fat as the amounts you eat will influence cholesterol levels in your blood. Note that *all fats* are high in calories. Avoid fried and fatty foods (remember there are hidden fats in many foods (e.g. meats, dairy products, nuts, cakes, biscuits, chocolates) and be sparing with all cooking and spreading fats. Wherever possible, choose low-fat alternatives.

Use polyunsaturated fats in preference to saturated fats. The type of fat you choose in your diet will affect the level of cholesterol in your blood. Saturated fats (mainly from dairy and animal sources) tend to raise blood cholesterol while polyunsaturated fats (from some vegetable and fish oils) can help to lower it. Therefore, it is beneficial to choose polyunsaturated fats (e.g. sunflower, corn, or soya oil/margarine) in preference to saturated types (e.g. butter, margarine, lard, or dripping).

NB Not all vegetable fats are polyunsaturated. Choose vegetable oil or margarine labelled high in polyunsaturates. Remember *all* fats, whether polyunsaturated or saturated, should be used sparingly.

Monounsaturated fats. These are also liquid at room temperature, the best known being olive-oil. This type of fat may be included in a low-fat diet in moderation.

Eat more fibre-containing foods. A high-fibre diet is known to be beneficial to health. Some types of fibre, commonly called soluble fibre are especially useful in helping to lower blood cholesterol. These are found in legumes and pulses (dried peas, beans, lentils) as pectins in fruits, and in porridge oats and oatmeal. Ensure you have plenty of fibre in your diet by including wholegrain cereals, wholemeal bread, plenty of vegetables and fruit, and regularly use pulses such as haricot beans, red kidney beans, and butter beans to name but a few of the very wide range available.

Cholesterol. While it is important not to take excessive amounts of cholesterol in food, it is far more important to watch the fat content of your diet. Eating too much saturated fat will cause far more harm than a small intake of cholesterol-rich foods. Therefore, it is more essential to take an overall low-fat diet than simply a low-cholesterol diet.

Sugar. Some foods which are high in fat (e.g. chocolate, biscuits, cakes, puddings, sweet pastries) also contain sugar and are high in calories. These should be avoided. Other sugary foods (e.g. jam, marmalade, honey) need to be reduced if you are trying to lose weight or if triglycerides are raised.

Alcohol. All forms of alcohol are high in calories and should be restricted if you are overweight. In addition, there is some evidence that alcohol may raise triglyceride levels. No one should drink excessively, but some people with hyperlipidaemia need to restrict alcohol consumption.

Appendix I Lipid metabolism: expanded notes

Apoproteins

These are the lipid-free protein components of the plasma lipoproteins. In general they are involved in maintaining structural integrity of the lipoprotein particles and they have a role in receptor recognition and enzyme regulation. They are classed A, B, C, D, and E with subclasses. Apoprotein A is the major protein in HDL. Apo A_1 binds phospholipid, activates lecithin cholesterol acyltransferase, and may have a function in regulation of membrane lipids and membrane fluidity. Apoprotein B constitutes 90 per cent of the protein of LDL and is a major protein of chylomicrons and VLDL. It is thought to have a vital function in the transport of triglycerides. Apoprotein CII activates the lipoprotein lipase of adipose tissue. Apoprotein E is involved with recognition of the remnant particle by the liver.

Fat absorption

Fats constitute 40 per cent of the calorie intake of many people. Partial hydrolysis of ingested fats occurs in the small intestine, due to the action of lipases, and in the presence of the bile salts, cholic and chenodeoxycholic acid, and some phospoholipids, micelles are formed. Monoglycerides and non-esterified fatty acids are absorbed in the duodenum and proximal jejunum, and are re-esterified in the endothelial cells to form triglycerides.

Dietary cholesterol esters are hydrolysed by pancreatic enzymes, and the cholesterol is absorbed in the small intestine. In the intestinal epithelial cells, triglycerides combine with cholesterol, phospholipids and specific apolipoproteins which have been absorbed or synthesized by the mucosal cells. The lipoproteins formed (chylomicrons and intestinal VLDL) are rich in triglyceride and are secreted into the lymphatic system where changes in cholesterol, phosophlipid and apoproteins occur, including loss of apo AII and uptake of apoproteins C and E.

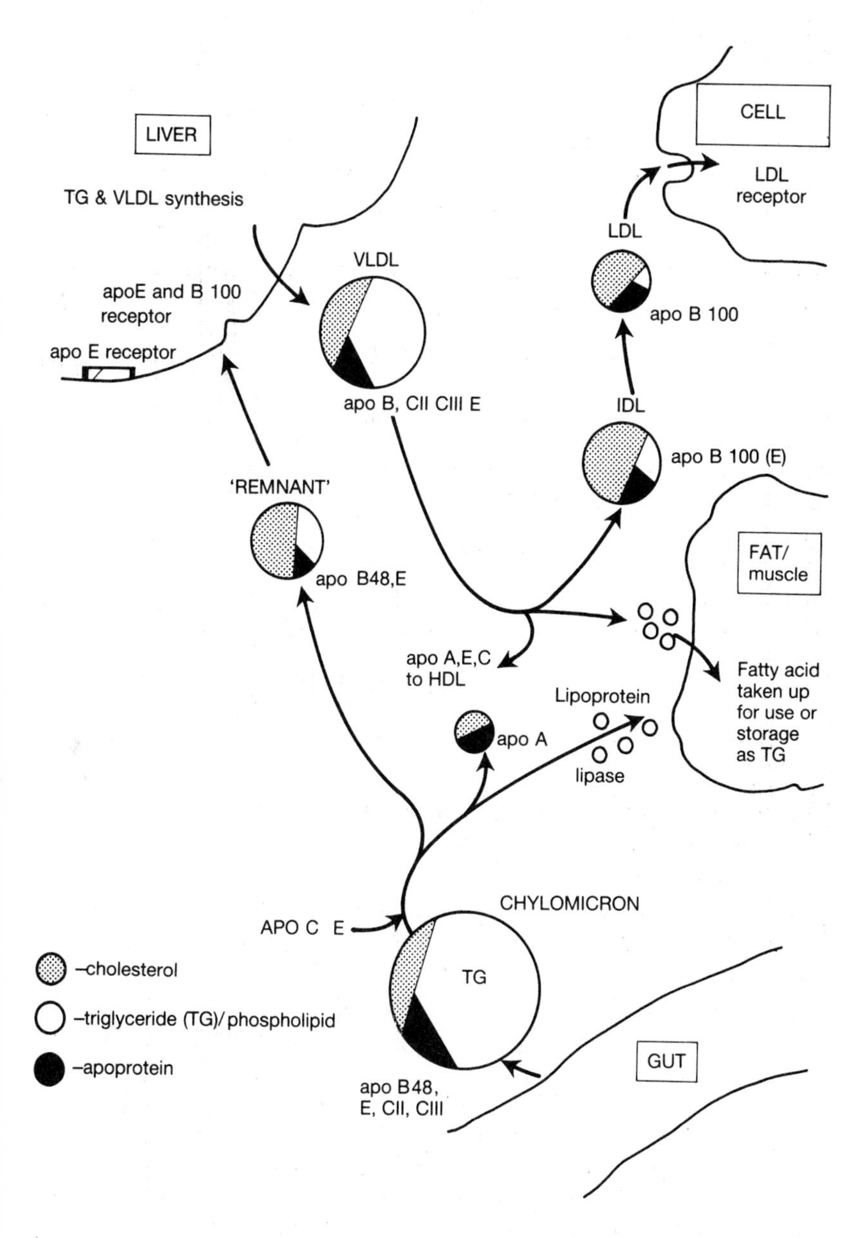

Fig. A.1. Pictorial summary of lipoprotein metabolism.

Fat transport

Chylomicrons enter the blood from the lymphatic system and may cause turbidity of the plasma after a fat-rich meal. Because triglycerides are insoluble in plasma they are transported as lipoproteins—macromolecular aggregates of variable size, lipid, and protein content. These lipoproteins are usually classified by density on ultracentrifugation. The lowest density lipoprotein with the greatest triglyceride content has the highest flotation number.

Lipoprotein flotation	Main lipid (Sf)	Main lipid	Main apoprotein
Chylomicrons	10^3–10^5	triglyceride	B
VLDL (very low density lipoprotein)	20–400	triglyceride	B,C,E
LDL (low density lipoprotein)	0–20	cholesterol	B
HDL (high density lipoprotein)		phospholipid	A,D

VLDL is also synthesized in the liver, and is the transport form of endogenously synthesized triglyceride.

In the circulation, triglyceride is gradually removed from the chylomicrons and VLDL, mainly by the action of lipoprotein lipase. Lipoprotein lipase is present in the capillaries of a number of tissues, but predominantly in adipose tissue, and it is stimulated by apoprotein CII present in the triglyceride-rich lipoprotein particles. Lipoprotein lipase is also stimulated by insulin and reduced activity occurs in poorly controlled diabetes.

Glycerides and non-esterified fatty acids removed by lipase action are taken up by muscle or adipose cells. These fatty acids provide the main energy source for aerobic metabolism in muscle and in a well-fed individual the excess is stored as triglyceride. As the triglycerides are removed the remnant particle becomes smaller and some of the more water soluble components on the surface become redundant. These include phospholipid, unesterified cholesterol, and the apo C molecules, which transfer to HDL. The metabolism of the chylomicron remnant or IDL is controversial. Some is probably taken up by the liver, and some is metabolized in the tissues to LDL. LDL is rich in cholesterol. Removal of LDL from the circulation is slower than that of many other particles. About half the LDL is initially bound by high-affinity cell receptors, and apoproteins B and E in the particle probably aid this binding. LDL then enters the cell and is degraded in lysosomes to liberate cholesterol which can be used by the cell. Dietary cholesterol inhibits endogenous cholesterol synthesis by inhibiting an enzyme in the pathway of cholesterol synthesis. The number of cell receptors appears to be regulated by the intracellular cholesterol level. There is also a low-affinity receptor pathway, which is not known to be regulated, but which becomes proportionally more important at higher LDL concentration. The activity of the high-affinity receptor is a major determinant of plasma LDL and cholesterol levels, and disorders affecting these receptors, such as occur in familial hypercholesterolaemia, have marked effects upon cholesterol metabolism and circulating plasma levels.

The other group of lipoproteins in the circulation are HDL_1, HDL_2, and HDL_3. These are mainly synthesized in the intestinal mucosa and liver. Phospholipids and cholesterol which become redundant when triglyceride-rich lipoproteins are metabolized are transferred to these particles, where cholesterol esters are formed by the action of lecithin cholesterol acyl transferase. These cholesterol esters may then transfer to other particles. The particular interest in this lipoprotein has arisen because the level of HDL_2 is inversely related to the risk of CHD.

Appendix 2: References and further reading

Epidemiological studies

Castelli, W.P. (1986). The triglyceride issue: A view from Framingham. *Am. Heart J.* **112**, 432–437.

Goldbourt, U., Holtzman, E., and Neufeld, H.N. (1985). Total and high density lipoprotein cholesterol in the serum and risk of mortality: evidence of a threshold effect. *Br. Med. J.* **290**, 1239–43.

Gordon, T., Kannel, W.B., Castelli, W.B., and Dawber, T.R. (1981). Lipoproteins, cardiovascular disease and death. The Framingham study. *Arch. Intern. Med.* **141**, 1128–31.

Meade, T., Mellow, S., Brozovic, M. *et al.* (1986). Haemostatic function and ischaemic heart disease: principal results of the Northwick Park Heart Study. *Lancet* **2**, 533–7.

Miller, N.E., Thelle, D.S., Forde, O.H., and Mjos, O.D. (1977). High-density lipoprotein and coronary heart disease: a prospective case-study. *Lancet* **1**, 965–7.

Pocock, S.J., Shaper, A.G., Phillips, A.N., Walker, M., and Whitehead, T.P. (1986). High density lipoprotein cholesterol is not a major risk factor for ischaemic heart disease in British men. *Br. Med. J.* **292**, 515–9.

Stone, N.J., Levy, R.I., Fredrickson, D.S., and Verter, J. (1974). Coronary artery disease in 116 kindred with familial type II hyperlipoproteinemia. *Circulation* **49**, 476–80.

Diet

Edington, J., Geekie, M., Carter, R., Benfield, L., Fisher, K., Ball, M., and Mann, J. (1987). Effect of dietary cholesterol on plasma cholesterol concentration in subjects following reduced fat, high fibre diet. *Br. Med. J.* **294**, 333–6.

Herold, P. and Kinsella, J. (1986). Fish oil consumption and decreased risk of cardiovascular disease: comparison of findings from animal and human feeding trials. *Am. J. Clin. Nutr.* **43**, 566–98.

Keys, A. (1970). Coronary heart disease in seven countries. *Circulation* **41**, 1–198.

Keys, A., Anderson, J.T., and Grande, F. (1965). Serum cholesterol response to changes in diet. 11 The effect of cholesterol in the diet. *Metabolism* **14**, 759–65.

Mann, J. and the Oxford Dietetic Group. (1982). *The diabetics' diet book: a new high-fibre eating programme*, Martin Dunitz.

Mattson, F.H. and Grundy, S.M. (1985). Comparison of effects of dietary saturated, monounsaturated, and polyunsaturated fatty acids on plasma lipids and lipoproteins in man. *J. Lipid Res.* **19**, 194–202.

National Advisory Committee on Nutrition Education. (1983). *Proposals for nutritional guidelines for health education in Britain*. Health Education Council, London.

Report of the Committee on Medical Aspects of Food Policy. (1984). *Diet and cardiovascular disease*. DHSS report, HMSO, London.

Royal College of Physicians of London. (1980). *Report on medical aspects of dietary fibre*. Pitman Medical, London.

Thorogood, M., Carter, R., Benfield, L., McPherson, K., and Mann, J.I. (1987). Plasma lipids and lipoprotein cholesterol concentrations in people with different diets in Britain. *Brit. Med. J.* **295**, 351–3.

Clinical trials—intervention

Blankenhorn, D. *et al.* (1987). Beneficial effects of combined colestipol-niacin therapy on coronary atherosclerosis and coronary venous bypass of grafts. *JAMA* **257**, 3233–40.

Brensike, J.F., Levy, R.I., Kelsey, S.F. *et al.* (1984). Effects of therapy with cholestyramine on progression of coronary atherosclerosis; results of the NHLBI type 11 coronary intervention study. *Circulation* **69**, 313–24.

Dayton, S., Pearce, M.L., Hashimoto, S., Dixon, W.J., and Tomiyasu, U. (1969). A controlled clinical trial of a diet high in unsaturated fat in preventing complications of atherosclerosis. *Circulation* **40** (suppl. 11), 58–60.

Duffield, R.G.M., Miller, N.E., Brunt, J.N.H., Lewis, B., Jamieson, C.W., and Colchester, A.C.F. (1983). Treatment of hyperlipidaemia retards progression of symptomatic femoral atherosclerosis. *Lancet* **2**, 639–42.

Helsinki Heart Study. (1987). Primary prevention trial with gemfibrozil in middle-aged men with dyslipidemia. *N. Engl. J. Med.* **317**, 1237–45.

Lipid Research Clinics Coronary Primary Prevention Trial Results. (1984). *JAMA* **251**, 351–75.

Multiple Risk Factor Intervention Trial Research Group. (1982). Multiple Risk Factor Intervention Trial: Risk factor changes and mortality results. *JAMA* **248**, 1465–77.

Nikkila, E.A., Viikiukoski, P., Valle, M., and Frick, M.H. (1984). Prevention of progression of coronary atherosclerosis by treatment of hyperlipidaemia: a seven year prospective angiographic study. *Br. Med. J.* **289**, 220–3.

Peto, R., Yusuf, S., and Collins, R. (1987). Cholesterol-lowering trial results in their epidemiological context. *Circulation* **75** (suppl. 2), 451.

WHO. (1980). Co-operative trial on primary prevention of ischaemic heart disease using clofibrate to lower serum cholesterol—mortality follow-up. *Lancet* **ii**, 379–85.

WHO European Collaborative Group. (1986). European collaborative trial of multifactorial prevention of coronary heart disease. *Lancet* **1**, 869–72.

Articles

Anggard, E. (1986). Prevention of cardiovascular disease in general practice: a proposed model. *Brit. Med. J.* **293**, 177–80.

Brown, M.S. and Goldstein, J.L. (1984). How LDL receptors influence cholesterol and atherosclerosis. *Sci. Am.* **250**, 58–66.

Consensus Conference. (1985). Lowering blood cholesterol to prevent heart disease. *JAMA* **253**, 2080–6.

Drugs and therapeutics bulletin. (1987). Drugs used for hyperlipidaemia.

Greten, H. (1987). Secondary hyperlipidaemias. *Lipid Review* **1**, 45–48.

Hoeg, J.M., Gregg, R.E., and Brewer, H.B. (1986). An approach to the management of hyperlipoproteinemia. *JAMA* **255**, 512–22.

Lewis, B. (1983). The lipoproteins: predictors, protectors, and pathogens. *Brit. Med. J.* **287**, 1161–3.

Lewis, B., Mann, J.I., and Mancini, M. (1986). Reducing the risks of coronary heart disease in individuals and in the population. *Lancet* **1**, 956–9.

Oliver, M.F. (1984). Hypercholesterolaemia and coronary heart disease: an answer. *Brit. Med. J.* **288**, 423–4.

Royal College of Physicians of London. (1983). Obesity. *J. Roy. Coll. Phys. Lond.* **17**, 5–65.

Shepherd, J., Packard, C.J., Bicker, S. *et al.* (1980). Cholestyramine promotes receptor-mediated low-density-lipoprotein catabolism. *N. Engl. J. Med.* **302**, 1219–22.
J. Med. **302**, 1219–22.
Slack, J. (1969). Risk of ischaemic heart disease in familial hyperlipoproteinaemic states. *Lancet* **2**, 1380–2.

Study Group, European Atherosclerosis Society. (1987). Strategies for the prevention of coronary heart disease: A policy statement of the European Atherosclerosis Society. *Eur. Heart J.* **8**, 77–88.

The British Cardiac Society Working Group on Coronary Prevention. (1987). Conclusions and recommendations. *Brit. Heart J.* **57**, 188–9.

United States Department of Health and Human Services. (1983). *The health consequences of smoking: cardiovascular diseases. A report of the Surgeon General.* US Department of Health and Human Services, Public Health Service. DHHS (PHS) 84–50204.

Index